Junkers Ju 8

and its Variants in World War II

Helmut Erfurth

Schiffer Military History
Atglen, PA

Translated by Dr. Edward Force

Book Design by Ian Robertson.

Copyright © 2002 by Schiffer Publishing.
Library of Congress Catalog Number: 2002111328

This work was originally published under the title *Vom Original zum Modell: Junkers Ju 88* by Bernard & Graefe Verlag.

Printed in China.
ISBN: 0-7643-1673-7

We are interested in hearing from authors with book ideas on related topics.

Published by Schiffer Publishing Ltd.
4880 Lower Valley Road
Atglen, PA 19310
Phone: (610) 593-1777
FAX: (610) 593-2002
E-mail: Schifferbk@aol.com.
Visit our web site at: www.schifferbooks.com
Please write for a free catalog.
This book may be purchased from the publisher.
Please include $3.95 postage.
Try your bookstore first.

In Europe, Schiffer books are distributed by:
Bushwood Books
6 Marksbury Avenue
Kew Gardens
Surrey TW9 4JF
England
Phone: 44 (0) 20 8392-8585
FAX: 44 (0) 20 8392-9876
E-mail: Bushwd@aol.com.
Free postage in the UK. Europe: air mail at cost.
Try your bookstore first.

Contents

Foreword

The Junkers Ju 88 is one of the most famous bombers of the Second World War. Originally conceived as a twin-engined high-speed bomber in 1935-36, the aircraft underwent numerous design changes in the years that followed. More than 3,000 improvements or changes were made to the basic design in the years between 1937 and 1944, resulting in more than 60 variants and six major production versions, of which the Ju 88 A-4 was built in by far the largest numbers.

With 15,100 examples produced, the Ju 88 represented one of the largest German procurement programs of the Second World War. The aircraft saw service over every front, including: operating as a long-range reconnaissance aircraft in the far north; as a dive bomber attacking European cities; a night fighter; long-range fighter; low-level attack aircraft; special transport; tank destroyer; a torpedo aircraft over the North Sea, the Atlantic and the Mediterranean; as

the bottom half of a *Mistel* stand-off weapon for attacks on bridges; and in a variety of roles over the expanses of the western deserts of North Africa.

The Junkers Ju 88 program was an outstanding achievement by the German aircraft industry in the areas of technology, design, and organization; however, it also was an integral part of the strategy of conquest by force of arms which engulfed Europe and a large part of the world. Seen in this way, technology, including this aircraft, had two faces. On the one hand the Ju 88 represented technical know-how and high-performance, with world records in speed and payload, and was well-liked by its crews. On the other hand, this aircraft was designed as a weapon of war and wrought havoc, whether destroying people or cultural treasures. And so the aircraft was also an instrument of a political ideology which brought death, destruction, and suffering to the European continent.

One of the first project drawings of the Ju 88 high-speed bomber.

The Birth of a Legend:
The Development History of the Ju 88

On 6 December 1933 Heinrich Koppenberg was assigned by the *Reichsluftfahrtministerium* (State Ministry of Aviation) to the board of directors of Junkers' Dessau facility, marking the start of a new phase in research and development. This reorganization in Dessau may be seen as a significant step in the formation and expansion of the new *Luftwaffe*. Heinrich Koppenberg was an ideal manager in every respect. For one thing, he was able to convince the labor force at Junkers that Hugo Junkers had retired voluntarily, making Koppenberg partly responsible for creation of this fable. The result was the misuse of the name of Junkers, the respected scientist and aviation pioneer, as a front for German air armament. With state support, Koppenberg succeeded in turning the Junkers factories into model National-Socialist operations. He refrained, however, from depriving the base plant in Dessau, especially the airframe and engine departments, of its status as a research facility. Instead, he did everything he could to gradually develop it into a center of technical innovation in Central Germany. Nearly a dozen operations in the Sachsen-Anhalt region were incorporated into the base plant. As a result, the number of employees was increased from 3,098 (in December 1933) to 11,613 in a relatively short time (in 1934). At the same time, the advances in research made by Junkers in the field of aircraft and engine construction were made available as licenses to the rest of the aviation industry, which made possible the profitable building of aircraft under license, thus saving the high costs of research associated with the development of a new design.

Ernst Zindel, chief aircraft designer at Junkers since 1923 and the creator of the Junkers Ju 52/3m, described this period (1934-35):

In 1935, following the development of the first aircraft for the new Luftwaffe, the Junkers Ju 86, Heinkel He 111, the Junkers Ju 87 dive bomber and other types, we at Junkers wondered what the next step would be. The 1935 edition of the Luftwaffe service manual Luftkriegführung 16 said: "As the offensive instrument of aerial warfare, the air forces, especially the bomber forces, define the nature of the air force." This was a precise and unmistakable definition of the air force's role in a future war. Consideration was already being given to developing the Junkers Ju 86 in various direc-

Ju 88 model in the wind tunnel.

tions to serve as a medium bomber. The result was to be a twin-engined high-speed bomber capable of carrying a number of bombs up to 50 kg in weight and also serving in the dive-bombing role. Project study and design work under the direction of Dipl.-Ing. August Wilhelm Quick was already in high gear when the aviatrix Marga von Etzdorf returned to Germany after examining the aviation industry in the United States. She reported that the Glenn L. Martin company had developed a fast twin-engined bomber which was claimed to possess a maximum speed of 450 km/h. This was an impressive performance at that time. By comparison the Junkers Ju 86 reached approximately 310 km/h with the Jumo 206 and the Heinkel He 111 approximately 330 km/h with the BMW VI or 380 km/h with the Jumo 211. Under the existing conditions the Americans had made a major and impressive advance by achieving such speeds.

Understandably, this news had an alarming effect (in Germany), and so the design bureau headed by Quick was assigned the task of designing a high-speed bomber with a maximum speed of at least 500 km/h. Chief designer Ernst Zindel's papers reveal that work was already proceeding on new designs in August 1935. The projected aircraft was to carry a medium bomb load of 800 to 1000 kg, accommodated within the fuselage to reduce drag, and achieve the desired maximum speed of at least 500 km/h.

The correctness of Junkers' projections was confirmed when, in autumn 1935, the *Reichsluftfahrtministerium* issued a specification to the aviation industry calling for a high-speed bomber capable of speeds approaching those of contemporary fighters. Of course, this was in keeping with existing concepts and planning for future bombers. The tactical concept was based on a bomber with much greater range and only slightly inferior maximum speed compared to the more heavily armed fighter aircraft. Its high speed would enable it to evade fighter attack, or make it impossible for attacking fighters to carry out more than one firing pass.

In autumn 1935, therefore, the Junkers designers were able to present to Director Koppenberg a follow-on development program describing what was to follow the Ju 86 and Ju 87. The proposal called for the development of a high-speed bomber followed by a high-altitude high-speed bomber, the later Ju 88, or a high-speed bomber or high-altitude reconnaissance aircraft capable of operating at altitudes of 9000 to 10000 meters at high speed. Junkers already had experience in the latter field, for Hugo Junkers had carried out work on a high-altitude aircraft with a pressurized cabin in 1928. The aircraft, the Junkers Ju 49, was equipped with an insulated pressure-tight, double-walled pressurized cabin able to accommodate a crew of two. The Ju 49 was powered by a Junkers L 88 engine producing 800 H.P. Junkers had also conducted research in the field and held various associated patents. The Junkers Ju 49 research aircraft was an advanced design for its day, especially when one considers that as late as 1935 it was able

The Ju 88 design team. On the right is chief designer Ernst Zindel, center Brunolf Baade.

to raise the world altitude record to approximately 13000 meters. One problem, however, was the lack of sufficiently-powerful engines for operations at extreme altitudes, a problem to which Prof. Mader dedicated himself.

At that time logistics played a major role, for the development and testing plans of the *Reichsluftfahrtministerium* specified an interval of about three years from the start of design work to completion of the first production aircraft. This development stage included preliminary planning, construction of the mock-up, acceptance and production approval by the *Reichsluftfahrtministerium*, design and construction of the first prototypes, factory trials, and performance and operational testing by the testing and development center in Rechlin. This would be followed by finalization of the production version, as well as production equipment and armament. Lessons learned in testing would then lead to refinements and a revision of the series with the associated modification and correction of construction blueprints. That was the design and experimental-practical part of the exercise. The technological process would then be finalized with the manufacture of jigs and tools for large-scale production. Associated with this would be the construction of a so-called "Zero Series" for the purpose of testing the jigs and tools for the later production version. These prototypes (such as the Ju 88 V1, for example) would subsequently be used for operational trials at the *E-Stelle Rechlin* or for further development, after which they would be assigned to front-line units as "Zero Series Aircraft" for service trials. Suggestions arising from the trials were considered from a design standpoint and incorporated into the overall technological process. It was at this point that the real problem arose, namely procuring materials for the production variant and initiating large-scale production.

The designers, technologists, and material planners in the Junkers factories could draw upon a wealth of experience in the field of logistics, however, completing this development work in a period of just three years was a fantastic technical and organizational feat for that time. A time reduction was only possible through the use of network planning, which the Junkers factories had been using successfully for some time. Time savings were achieved by overlapping testing and series preparation, and by "telescoping" design and production technologies. Of course, such

Design drawing for the Ju 88 A-series.

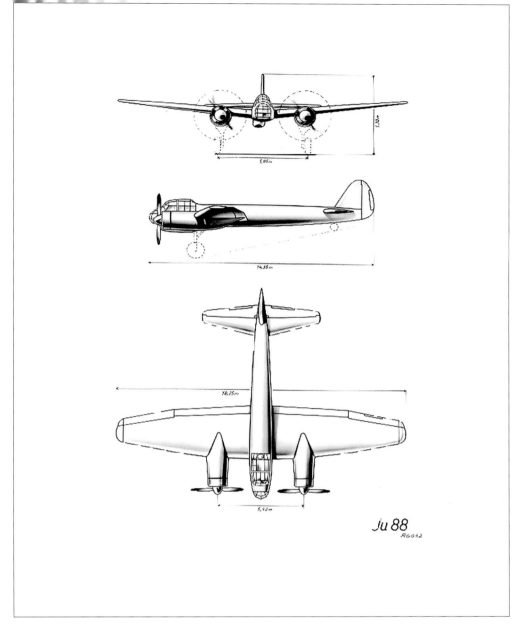

Ju 88
R6012

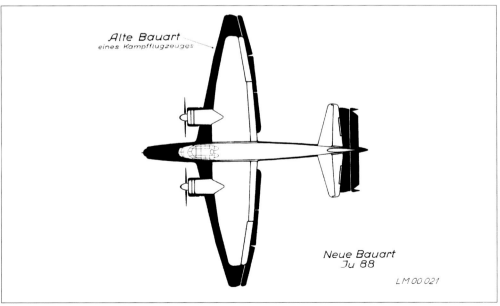

Alte Bauart
eines Kampfflugzeuges

Neue Bauart
Ju 88

LM 00 021

Comparison of the Ju 86 and Ju 88 bombers.

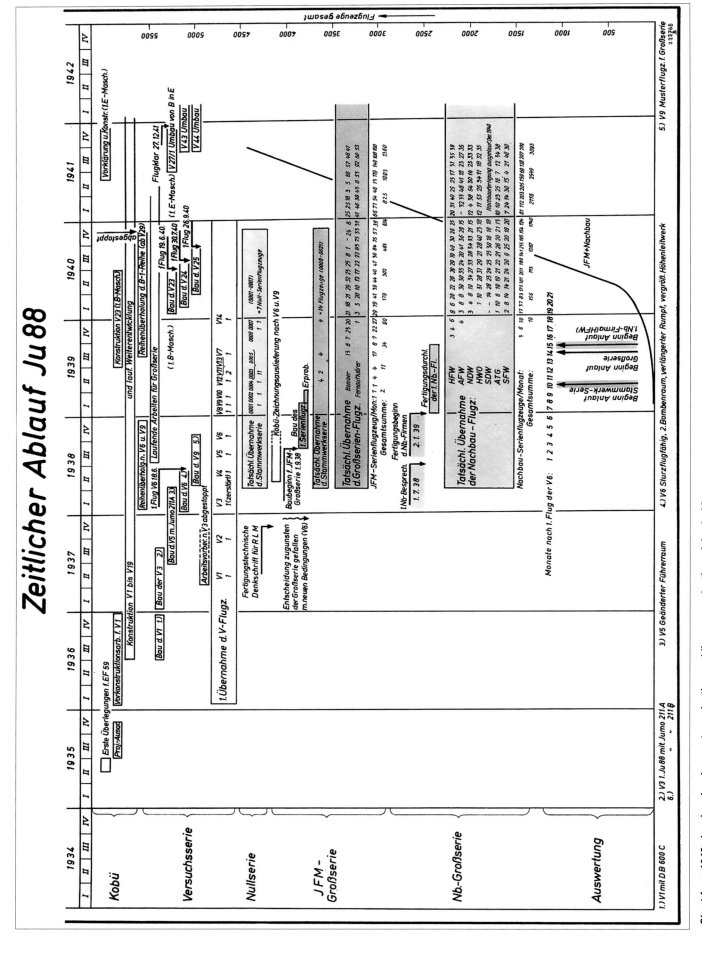

Chart from 1942 showing development, production and license production of the Ju 88.

Vergleich der Abmessungen und Gewichte

Type	Ju 52 K (BMW 132 A)	Ju 86 K (BMW 132 Dc)	Ju 88 (Jumo 211 B)
Spannweite	29,25 m	22,50 m	18,25 m
Länge	18,90 m	17,87 m	14,35 m
Höhe	6,10 m	4,70 m	5,30 m
Flügelfläche	110,5 m²	82,00 m²	52,50 m²
Rüstgewicht	6470 kg	5970 kg	7250 kg
Zuladung	4030 kg	2530 kg	3110 kg
Fluggewicht	10500 kg	8500 kg	10367 kg
Landegewicht	10500 kg	8500 kg	9350 kg
Startleistung	3×660 = 1980 PS	2 × 850 = 1700 PS	2×1220 = 2440 PS

Table of dimensions and weights of various aircraft types.

optimization was accompanied by an element of risk. In this situation a proposal by Hans Wagner, who had joined the Junkers factory in Dessau and to whom Koppenberg had assigned primary design responsibilities, came at the right moment. Wagner, who had been a professor in Danzig and Berlin, was an outstanding theoretician who had come to the aviation industry to assist in the development and testing of aircraft. At a conference attended by Koppenberg and Zindel, he proposed simply skipping over the planned development of a high-speed bomber with an optimal operating altitude of about 5000 meters and proceeding immediately with the development of a high-altitude, high-speed bomber with a pressurized cockpit. The result was the EF 61, which the *Reichsluftfahrtministerium* had ordered as an experimental high-altitude bomber, but with no intention of placing it in production. For its part, the *Reichsluftfahrtministerium* pressed for the planned Ju 88 high-speed bomber, which was supposed to replace the Heinkel He 111. It was understandable, therefore, that Baron von Richthofen, then the director of the *C-Amt* (the Technical Department of the *Reichsluftfahrtministerium*'s Technical Office), personally placed Zindel in charge of the Ju 88's development in the presence of Koppenberg. At that time the Junkers Dessau facility was working on a number of other projects, including the Ju 87 dive bomber and the Ju 89 long-range bomber. A planned civil variant of the latter, the Ju 90, and the Ju 85 light bomber were abandoned in favor of the Ju 88.

In January 1936 a situation arose in Ernst Zindel's design bureau which has led to speculation in some aviation literature. Through the intervention of General Director Koppenberg and following a visit to the United States by Dr. Hans-Heinz Hagemann of Junkers Dessau, two young engineers, Wilhelm Heinrich Evers and Alfred Gassner, were hired. Both engineers had experience in light metal stressed-skin construction. The pair had worked for the aircraft manufacturer Fairchild in Hagerstown, where they designed small aircraft. Junkers sent them to Dessau in late December 1935 and made them responsible for completing the design work on time. After the *Reichsluftfahrtministerium* again confirmed the role of the future bomber, work began on the aircraft's structural and aerodynamic form. Airflow tests were conducted in a wind tunnel using wooden models. The new smooth metal skin, used by Junkers since 1934, offered significant aerodynamic advantages. Corrugated metal was replaced by sheet duralumin for wings and fuselages. The result was lower drag and an additional increase in speed. The Ju 86 was the first Junkers aircraft to have a retractable undercarriage. As development progressed and the cantilever wing became thinner, with decreased depth and increased wing loading, a change was made from the old steel tube construction with riveted diagonal struts to the so-called "flat flange bearer" with continuous metal stiffeners to achieve the maximum and equal use of the flange material. Developed by Zindel, this mode of construction was first used in the Ju 87 in 1935 and was also chosen for the Ju 88.

At Christmas 1936, shortly before the maiden flight of the Ju 88 V1, Alfred Gassner and his wife visited relatives in the United States and did not return to Germany. The reason for his decision remains a mystery, however, Gassner allegedly believed that he was under constant surveillance by the Gestapo.

After Gassner's departure, Wilhelm Heinrich Evers went on alone, working on the development, construction, and testing of the V2 to V5 prototypes. One important area of effort was the change from the DB 600 engine to the Jumo 211 VA beginning with the V3. In addition, he also had to deal with the constant requests for changes from the RLM concerning the aircraft's equipment and armament.

In the summer of 1938 Brunolf Baade took over development of the Ju 88, while Evers assumed responsibility for the introduction of the type into large-scale production. Evers remained in Dessau and Bernburg until 1945, continuing in his role as coordinator. From 1939 designers Karl-Ernst Schilling, and Bernhard Cruse and their design groups worked to develop the Ju 88 series, however, overall responsibility for the Ju 88 from initial concept to service introduction remained in the hands of chief designer Ernst Zindel.

A *Furor Teutonicus* in Stressed-Skin Construction: The Design of the Ju 88

The Junkers Ju 88 was an aircraft whose basic design was modified to fulfill a wide variety of roles—conventional and dive bomber, marine and strategic reconnaissance aircraft, heavy fighter, night fighter, and tank destroyer. The aircraft was also used as a so-called "Bottom Aircraft" in the *Mistel* combination. Like almost no other type operated by the *Luftwaffe*, it embodied the latest developments in aviation technology, as well as practical improvements resulting from flight testing and front-line service.

Technical guidelines issued by the *Reichsluftfahrtministerium* for the construction of a high-speed bomber in October 1935 marked the starting point in the design of the Ju 88. These called for a twin-engined conventional bomber with a crew of three. The required performance was:

Maximum speed: 500 km/h
Minimum range: 2000 km
Bomb Load: 500 kg, consisting of ten 50-kg bombs accommodated in the fuselage.

Armament: one fixed MG 17 in the fuselage nose and one flexibly-mounted MG 15 for rear defense.

Junkers submitted its proposal to the *Reichsluftfahrtministerium* in December 1935, and design work on the aircraft took place in 1936. The wing—trapezoid shape in outline—was mounted low on the fuselage. It was of stressed-skin monocoque construction, which was new to Junkers at that time. Instead of the old tubular spar construction, a "flat flange bearer design" with continuous metal stiffeners was used. This resulted in a more rational use of materials, which also meant that the design's aerodynamic loads were more evenly distributed. This factor was of particular importance for the dive-bombing role. Annular radiators mounted in front of the engines were selected because of their lower drag. An old idea of Professor Junkers, Herbert Wagner had used front-mounted radiators on the EF 61. In an effort to minimize drag, the bombs were to be carried in cells within the fuselage.

To meet the requirements of large-scale production, the Ju 88's fuselage was built as upper and lower shells. This method of construction had the advantage of dramatically accelerating assembly of the fuselage, as it was possible to work on the inside and outside of each fuselage half at the same time. This method, which will be described in detail, was a significant factor in the design of the Ju 88. It showed how Junkers had combined technology and organization in the design stage of the project.

Structurally, the fuselage consisted of four main stringers, four secondary stringers, and 33 bulkheads, which gave the fuselage shape and rigidity. Each fuselage half had two main stringers and two secondary stringers. In addition, two

Bottom fuselage shell under construction.

Top fuselage shell under construction.

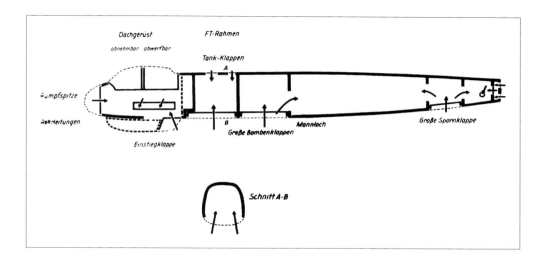

Access to the fuselage interior.

short stringer sections stiffened the rear fuselage in the area of the tailwheel cutout in order to more evenly distribute the forces acting on the fuselage there. The stringers consisted of special T-sections produced individually and shaped in special jigs so that they could be installed in the fuselage with no stress. After the fuselage halves were completed, the outer skin of duralumin sheet was riveted to the frame. The upper and lower fuselage halves were assembled in separate jigs, which made it possible to use semi-skilled labor in some areas. Technically and organizationally, this was an important aspect of the Ju 88 production line method.

Once the two halves were completed, they were assembled in a special jig by riveting the outer skin to all the bulkheads. This was followed by installation of the cockpit area and glazing. The framework for the nose transparency was of cast *Elektron*, while the main canopy framework was of welded steel tube.

The ventral gondola of the Ju 88 was a separate component and was attached to the fuselage with rivets. The folding part of the ventral gondola, which served as an entrance-exit hatch, also accommodated the mounting for the ventral machine-gun.

The Ju 88's wing consisted of two spars, on which ribs were arranged. The engine and undercarriage attachment points were also located on the spars, so that all of the main assemblies (airframe, wings, and the two engines) formed a statically stable arrangement. This self-contained, statically determined distribution of forces was necessary, as the aircraft was subjected to high loads, especially in the dive-bombing role. The solidity of the Ju 88's design made possible the multitude of variants derived from the basic design. Experiments aimed at developing a tank-destroyer saw the Ju 88 equipped with heavy cannon of various calibers. Firing these large-caliber weapons placed great stress on the fuselage, however, only minor deformations were found, and these were overcome by local strengthening.

The production dive bomber was equipped with dive brakes and an automatic pull-out system. All of the flight controls were located in the cockpit, which accommodated the pilot, bomb aimer, and gunners. The pilot used the control column to activate the elevator and ailerons, while a secondary control column provided the bomb aimer with elevator and aileron control. Supplementing the control system was a Siemens K 4 single-axis autopilot.

Ju 88 D wing with flaps and ailerons removed.

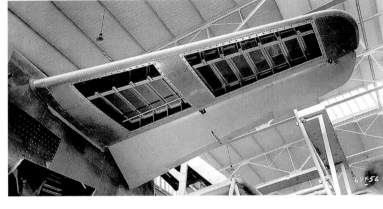

Horizontal tail of a Ju 88 A-4.

Fin and rudder of a Ju 88 A-4.

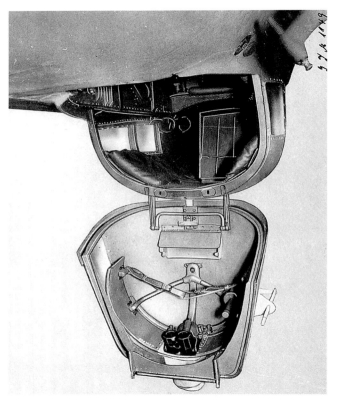

Ju 88 ventral gun position (*C-Stand*) with folding ventral gun mount/escape hatch.

Interior view of the fuselage of a Ju 88 A-4. Note the oxygen bottles in the aft fuselage.

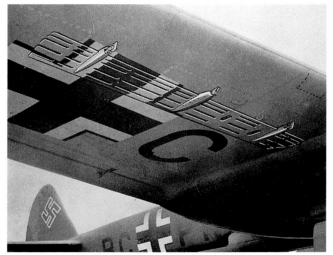

Retracted dive brake.

Dive brake in open position during assembly.

Ju 88 A-4 with ventral hatch (Bola) in open position.

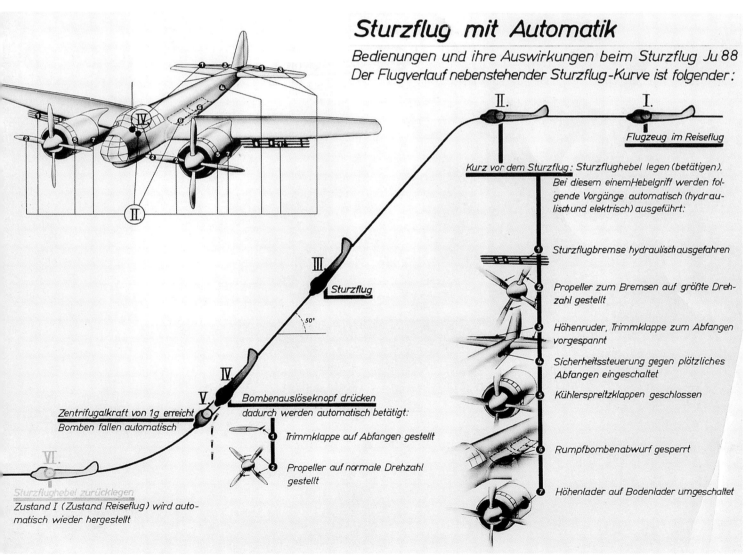

Sturzflug mit Automatik

Bedienungen und ihre Auswirkungen beim Sturzflug Ju 88
Der Flugverlauf nebenstehender Sturzflug-Kurve ist folgender:

II.   **I.**

Flugzeug im Reiseflug

Kurz vor dem Sturzflug: Sturzflughebel legen (betätigen).
Bei diesem einem Hebelgriff werden folgende Vorgänge automatisch (hydraulisch und elektrisch) ausgeführt:

1. Sturzflugbremse hydraulisch ausgefahren

2. Propeller zum Bremsen auf größte Drehzahl gestellt

3. Höhenruder, Trimmklappe zum Abfangen vorgespannt

4. Sicherheitssteuerung gegen plötzliches Abfangen eingeschaltet

5. Kühlerspreitzklappen geschlossen

6. Rumpfbombenabwurf gesperrt

7. Höhenlader auf Bodenlader umgeschaltet

III.
Sturzflug

50°

IV.

Zentrifugalkraft von 1g erreicht
Bomben fallen automatisch

Bombenauslöseknopf drücken
dadurch werden automatisch betätigt:

1. Trimmklappe auf Abfangen gestellt

2. Propeller auf normale Drehzahl gestellt

VI.
Sturzflughebel zurücklegen
Zustand I (Zustand Reiseflug) wird automatisch wieder hergestellt

V.

Drawing illustrating how the Ju 88's automatic pull-out mechanism functioned.

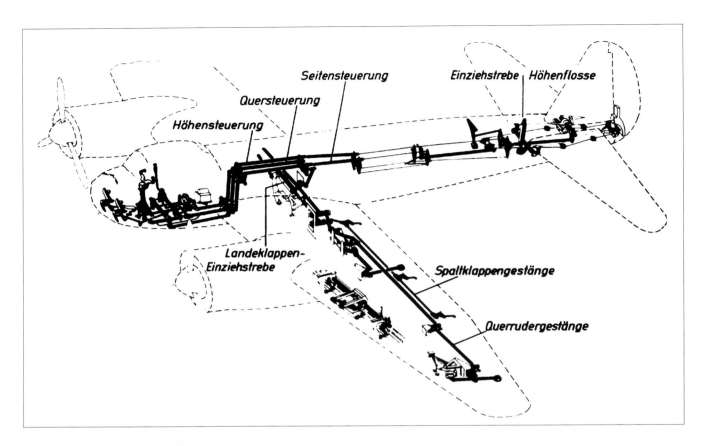

Schematic drawing of the Ju 88's control system.

In addition to the armament, which will be dealt with elsewhere, there were items of special equipment needed by the pilot or other members of the crew. The Ju 88 D-1 strategic reconnaissance aircraft had a camera installed in the fuselage behind Bulkhead 15. A so-called "Kärcher Stove" provided heat to prevent the system from freezing at high altitude. For operations at high altitude there were also de-icing systems for the propellers and wings and oxygen equipment for the crew. Tropical versions had additional water tanks and other items of equipment. Night fighter versions of the Ju 88 were equipped with externally-mounted radar antennae, which were dubbed "Antlers."

Port engine of a Ju 88 illustrating the annular radiator and spinner.

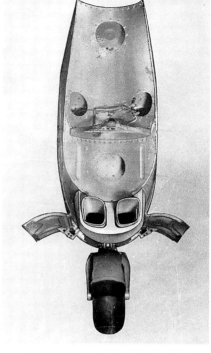

geschlossen geöffnet

LM00044

| JFM 1939 | Kühlerspreizklappen der Junkers-Ju 88 | JFM 145 |

Ju 88 radiator flaps in the closed and open positions.

Aft fuselage with tail skid and fuel jettison vents.

Below: Hot air and de-icing systems.

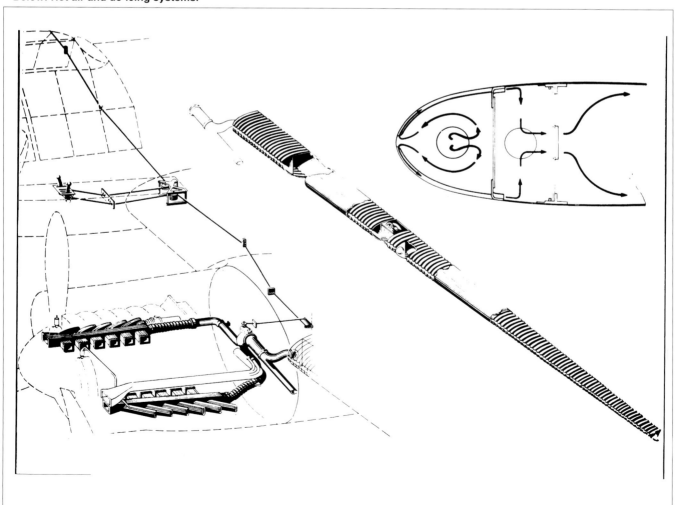

Ju 88 main undercarriage member in the extended position.

Ju 88 main undercarriage member during retraction sequence.

Mainwheel bay doors closed.

Tailwheel of a Ju 88 A-4/D-1 in the retracted position with doors open.

JUNKERS- JU 88 A-1

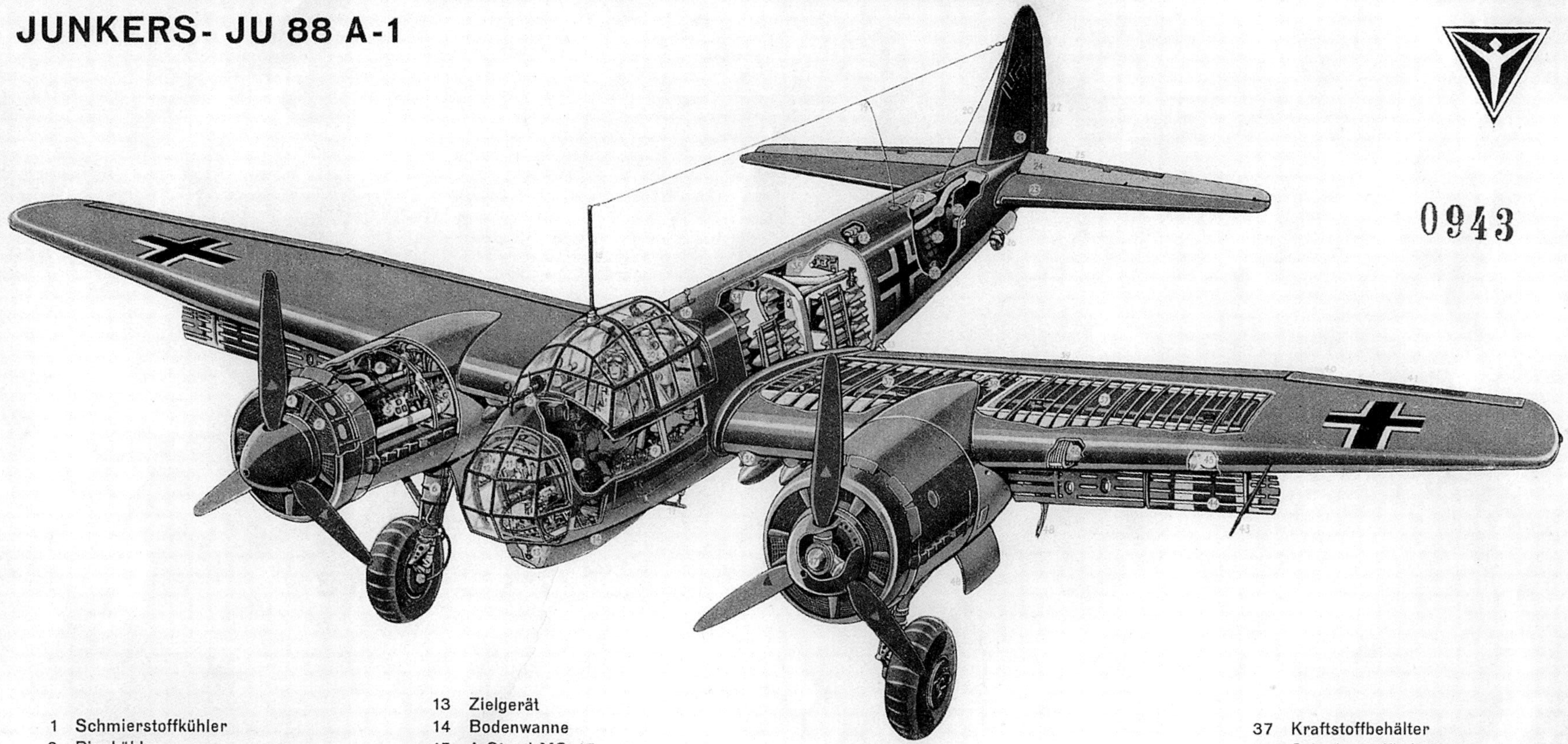

0943

1 Schmierstoffkühler
2 Ringkühler
3 Kühlerklappen
4 Motor Jumo 211
5 Anzeigegeräte
6 Einziehbares Fahrwerk
7 Führerraum
8 Steuersäule
9 Bedienanlage
10 Abwerfbares Führerraumdach
11 Leuchtpatronenkasten
12 RAB (Reihenabwurfgerät für Bomben)

13 Zielgerät
14 Bodenwanne
15 A-Stand MG. 15
16 B-Stand MG. 15
17 C-Stand MG. 15
18 Antennenmast
19 Antenne
20 Notantenne
21 Seitenflosse
22 Seitenruder mit Trimmklappe
23 Höhenflosse
24 Höhenruder
25 Trimmklappe für Höhenruder
26 Einziehbares Spornrad

27 Sanitätspack
28 Schlauchboot
29 Kraftstoff-Schnellablaß
30 Sauerstoff-Flaschen für Höhenatmer
31 Mutterkompaß
32 Peilgerät
33 Schleppantenne
34 Vorderer Bombenraum
35 Hinterer Bombenraum
36 Lastenträger

37 Kraftstoffbehälter
38 Schmierstoffbehälter
39 Landeklappe
40 Querruder
41 Trimmklappe für Querruder
42 Positionslampe (Backbord)
43 Staurohr
44 Sturzflugbremse (2-teilig)
45 Scheinwerfer
46 Enteisungsanlage
47 Verstelluftschraube
48 Fahrwerksklappen

Junkers Flugzeug- und Motorenwerke A.G., Dessau

Zchg. Schaffer

Lehrmittel-Abteilung LM-Nr. 562

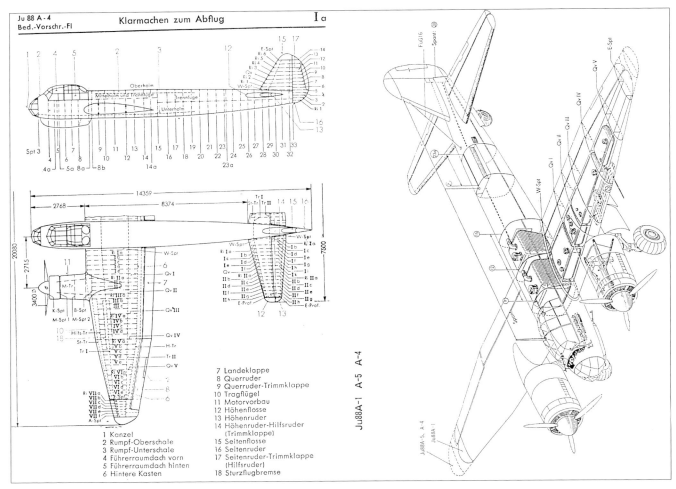

1 Kanzel
2 Rumpf-Oberschale
3 Rumpf-Unterschale
4 Führerraumdach vorn
5 Führerraumdach hinten
6 Hintere Kasten

7 Landeklappe
8 Querruder
9 Querruder-Trimmklappe
10 Tragflügel
11 Motorvorbau
12 Höhenflosse
13 Höhenruder
14 Höhenruder-Hilfsruder (Trimmklappe)
15 Seitenflosse
16 Seitenruder
17 Seitenruder-Trimmklappe (Hilfsruder)
18 Sturzflugbremse

Ju 88 design drawings.

Patent for jettisonable bodies on aircraft, especially jettisonable fuel tanks (drop tanks).

Patent for rapid emptying of aircraft fuel tanks.

18

From Prototype to Production Aircraft

The technological and organizational preparations for large-scale production went hand in hand with design development of the Ju 88. Experts were consulted as to how to eliminate economic bottlenecks in the materials economy; as well, there was a shortage of qualified workers. Significant delays were encountered in the procurement of raw materials, some of which were regulated by the *Reichsluftfahrtministerium* according to priority. For example, instead of 640 metric tons of duralumin, which would have been sufficient to cover Junkers' requirements for Ju 88 production for one month, the RLM only authorized 200 tons of the material. In the month of August 1939, only a few

Der Führer und
Reichskanzler

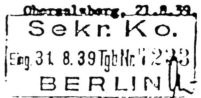

Obersalzberg, 21.8.39.
Sekr. Ko.
Eng. 31. 8. 39 Tgb.Nr. 7 2 3
BERLIN

Ich beauftrage den Reichsminister der Luftfahrt und Oberbefehlshaber der Luftwaffe bei der Aufrüstung der Luftwaffe das Flugzeugmuster Ju 88 und die 8,8 cm Flak besonders zu bevorzugen und die Ausbringung auf mindestens:

300	Ju 88 monatlich	(davon 50 als Reserveteile)	) inklusive
100	8,8 cm Flak monatlich		) der not-
30	10,5 cm Flak monatlich		) wendigen Munition

zu steigern.

Das hierfür erforderliche zusätzliche Material, Personal, die Werkzeugmaschinen und die zum Ausbau der Industrie benötigten Baumaterialien und Bauarbeiter sind mit der gleichen Dringlichkeit zur Verfügung zu stellen, wie ich . sie für den Aufbau der Kriegsmarine befohlen habe.

Der Beauftragte für den Vierjahresplan hat die weiteren Ausführungsbefehle zu erlassen.

Directive from Adolf Hitler to Hermann Göring, Reich Minister of Aviation and Commander-in-Chief of the *Luftwaffe*, dated 21 August 1939 ordering an increase in production of the Ju 88.

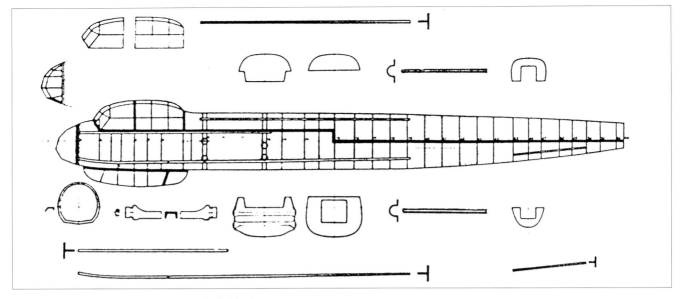

Breakdown of the Ju 88 fuselage into individual components.

weeks before the outbreak of war, Junkers recorded a shortfall of more than 30,000 metric tons of steel. There were also bottlenecks in certain types of materials, so that alternative solutions had to be sought. As well, components were not delivered on time because the subcontractors had not been sufficiently prepared. There were also problems concerning the quality of materials and the provision of semi-finished products, which were for the most part provided by subcontractors. Another problem area was the procurement of necessary machine tools, jigs, die blocks, tools, and production and measuring equipment.

A smooth-running production program required a central control and coordination office to organize the entire Ju 88 airframe and engine program. In terms of logistics, involving eight aircraft factories in the production of a single aircraft type was certainly a novelty. In 1938 and 1939 the manufacturer failed to deliver the required number of aircraft. Hermann Göring subsequently ordered the director of Junkers, Koppenberg, to personally take charge. The result was central control. In 1941 it was outlined in a 32-page document which was designated a state secret. The document accurately portrayed this type of coordination and control as a step towards rationalizing production in the aircraft industry.

Central control meant:

Supplying aircraft factories with tools, jigs, drawings, materials, parts, machines, etc.;
a modification service, for coordination of subsequent changes in production;
control of a total of 36 airframe and engine factories, including license companies, plus the approximately 3,500 subcontractors;
control of the flow of materials, the so-called bottleneck components and domestic equipment technology;
assistance in overcoming problems in production, especially relating to insufficient capacity by the subcontractors, through monitoring agencies, through so-called schedule trackers, back-up labor forces, training courses leading to qualification, etc.;

constant monitoring of production with emphasis on meeting schedules and technically problem-free deliveries, such as issuing a large number of individual parts to license companies or subcontractors when bottlenecks arise, in order to keep production flowing.

Junkers also submitted proposals to the RLM on how to meet the required program and eliminate bottlenecks. Besides the procurement of raw materials and the coordination of production, there was another problem: labor forces. The previously cited study stated that 200,000 people were required for the Ju 88 program. Special measures by Junkers enabled it to gradually reduce the previous shortfall of 35,000 workers. The study also reveals that the use of non-German labor, such as foreign workers, forced laborers, and prisoners of war was being considered from the very start of the war. It was also anticipated that production facilities in captured territory would be exploited, or that tools and complete engine factories would be transported to Germany as war booty. Following the incorporation of Austria into the Reich and the annexation of much of Czechoslovakia in 1938 and 1939, much of the production capacity in those areas was incorporated into the German armaments program. There were much more significant developments in 1940 following the German invasion of Western Europe. The following is a chronology of those events:

9 April 1940	occupation of Norway
14 April	a delegation from the *Reichsluftfahrtministerium* travels to Norway to assess the stocks of aluminum available there
18 April	first purchase and delivery agreement with Norwegian aluminum firms by Dr. Koppenberg
8 May	decree by the *Führer* that all existing stocks of aluminum and the existing aluminum production capacity in Norway is to be placed at the direct disposal of the *Luftwaffe*

Ju 88 fuselages under construction in the Junkers branch plant in Bernburg.

10 May	Dr. Koppenberg was appointed custodian of the Norwegian aluminum industry
23 July	Dr. Koppenberg given full authority for the aluminum industry of the occupied western territories in neutral Switzerland
20 August	formation of the Junkers Aluminum Office
1 October	Dr. Koppenberg appointed administrator of enemy property of the Norwegian aluminum industry
12 November	order by *Reichsmarschall* Göring for the removal of the Norwegian aluminum industry
3 December	formation of NORDAG—the Northern Aluminum Company.

These activities were continued in 1941:

30 April 1941	Koppenberg ordered to secure a raw materials base in the southeast area for the Norwegian aluminum industry
28 May	formation of the Hansa-Leichtmetall-AG.

These measures, which were carried out with almost military precision, enabled the RLM and Junkers to obtain the required aluminum and alumina, and regulate the necessary deliveries of bauxite to the benefit of the German air armaments industry. The goal was to produce one million tons of aluminum annually by 1944. These annexations and monopolizations of aluminum stocks or their production for the German armaments industry were vital, for as the war progressed shortages of materials became more commonplace and alternative solutions had to be found. The quality of materials also declined as the war progressed, and the situation was exacerbated by sabotage and various resistance movements.

With the support of those companies building the Ju 88 under license (Flugzeugwerke Arado in Brandenburg an der Havel, ATG in Leipzig, Dornier in Friedrichshafen, Heinkel in Oranienburg, Henschel in Schönefeld near Berlin, Dornier's factory in Wismar in northern Germany, Siebel in Halle, the Volkswagen automobile company in Fallersleben, and Opel in Rüsselsheim), the Junkers Flugzeugbau in

Installing the cockpit canopy.

Working on the ventral gondola.

22

Completed Ju 88s at the Junkers airfield in Bernburg awaiting final acceptance.

Ju 88 V 21 during winter trials at the Junkers airfield in Dessau, December 1940.

Dessau with its many branch operations in central Germany succeeded in achieving a significant increase in production of the Ju 88 in the period from August 1939 to March 1940. This decentralization of pre- and final assembly within Germany caused some problems in the area of transport, however, it later proved beneficial when the military situation changed and German centers were subjected to constant bombing by the Allies. The logistics of the work flow plan also proved extremely efficient. The complete decentralization of production and the subdivision into empennage, wing, fuselage, and final assembly using the assembly line method could be checked at any time and made it possible to eliminate limiting factors. This organization made it possible to increase production as the war went on, in spite of material shortages and frequent air attacks.

The original Junkers factory in Dessau was primarily engaged in the development, design, and testing of new prototypes. The Allies also viewed this part of the German aviation industry as principally a research facility, which it was. Following the departure of Hugo Junkers, the Nazis retained the plans for large-scale production which he had developed and refined them in the period that followed. Whereas finally assembly had previously been divided into six phases, it now consisted of several work places with stationary scaffoldings. A certain number of workers were always needed in each area, repeating the same task within a specified interval. This method was refined further to make production even better and more economical. Junkers planning and production engineers determined the individual production phases using network planning. Special newly-designed jigs, instruction stands, and test stands supplemented the work flows, so that large components and the product aircraft itself could be assembled over an assembly line. Each work station had a precise work plan with times calculated to the minute, the so-called "phase duration," plus a large number of various work processes which had to be carried out on each component. A new work group at the next station took over the component and continued work on it, moving it one step closer to completion. By concentrating the system it was possible to regulate the number of aircraft produced. This of course depended on requirements and the demands of the authorities that issued work orders. This way of working also guaranteed that certain refinements could be introduced into the system without halting production. Junkers manufactured both aircraft and engines using this system. After the concept employed by Junkers in Dessau proved successful, it was gradually adopted by other factories in the German aviation industry.

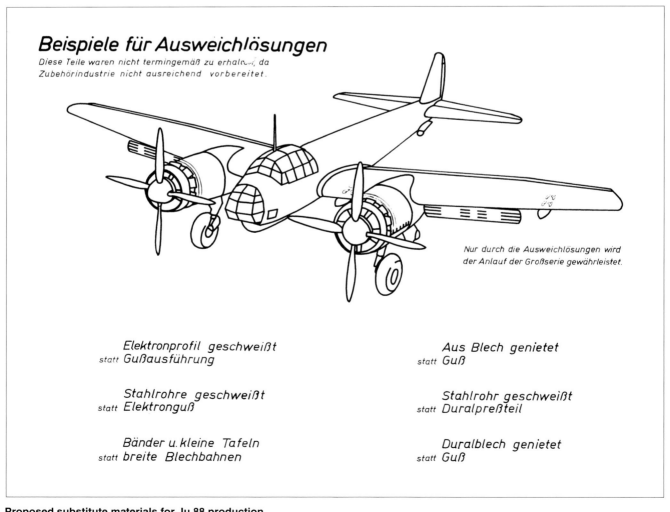

Proposed substitute materials for Ju 88 production.

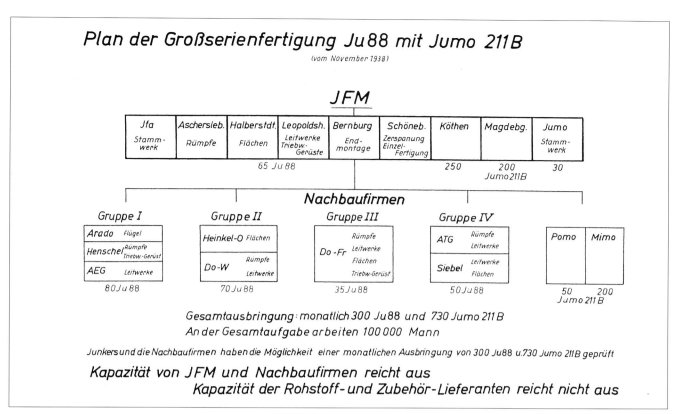

Plan der Großserienfertigung Ju 88 mit Jumo 211 B

(vom November 1938)

JFM

Jfa	Aschersleb.	Halberstdt.	Leopoldsh.	Bernburg	Schöneb.	Köthen	Magdebg.	Jumo
Stamm-werk	Rümpfe	Flächen	Leitwerke Triebw-Gerüste	End-montage	Zerspanung Einzel-Fertigung			Stamm-werk

65 Ju 88   250   200   30
                                Jumo 211 B

Nachbaufirmen

Gruppe I
Arado	Flügel
Henschel	Rümpfe Triebw-Gerüst
AEG	Leitwerke

80 Ju 88

Gruppe II
| Heinkel-O | Flächen |
| Do-W | Rümpfe Leitwerke |

70 Ju 88

Gruppe III
| Do-Fr | Rümpfe Leitwerke Flächen Triebw-Gerüst |

35 Ju 88

Gruppe IV
| ATG | Rümpfe Leitwerke |
| Siebel | Leitwerke Flächen |

50 Ju 88

| Pomo | Mimo |

50   200
Jumo 211 B

Gesamtausbringung: monatlich 300 Ju 88 und 730 Jumo 211 B
An der Gesamtaufgabe arbeiten 100 000 Mann

Junkers und die Nachbaufirmen haben die Möglichkeit einer monatlichen Ausbringung von 300 Ju 88 u. 730 Jumo 211 B geprüft

Kapazität von JFM und Nachbaufirmen reicht aus
Kapazität der Rohstoff- und Zubehör-Lieferanten reicht nicht aus

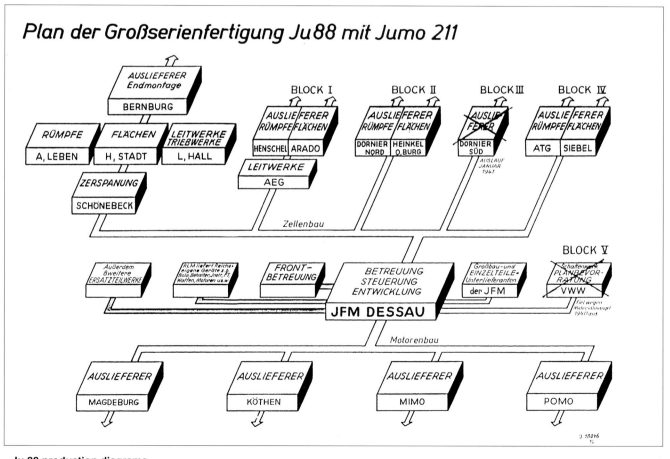

Plan der Großserienfertigung Ju 88 mit Jumo 211

Ju 88 production diagrams.

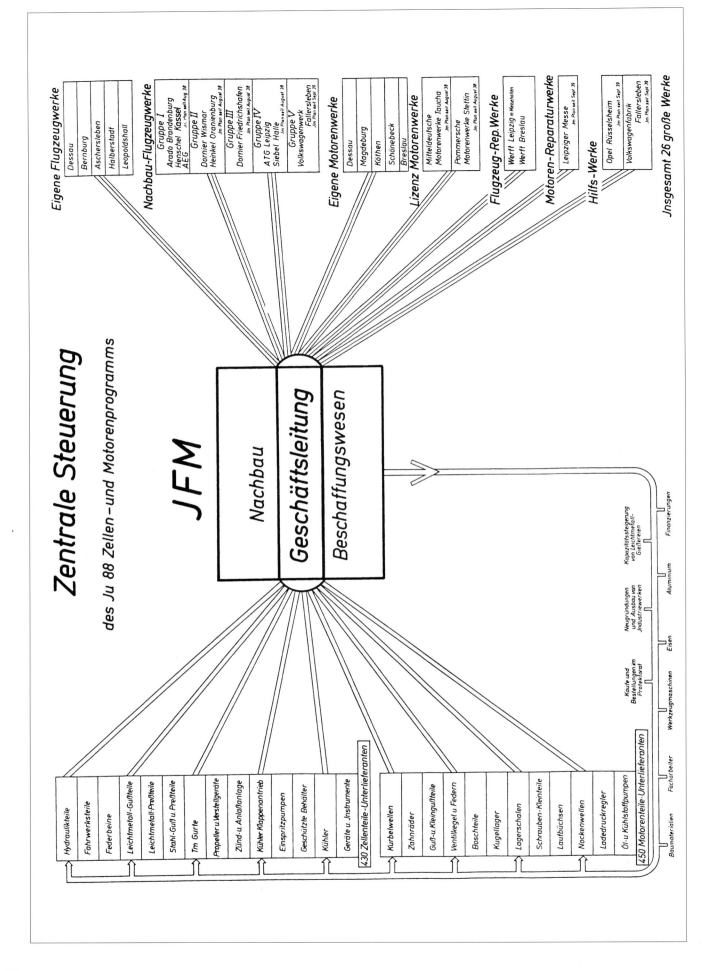

Zentrale Steuerung

des Ju 88 Zellen- und Motorenprogramms

JFM

Nachbau
Geschäftsleitung
Beschaffungswesen

Eigene Flugzeugwerke
- Dessau
- Bernburg
- Aschersleben
- Halberstadt
- Leopoldshall

Nachbau-Flugzeugwerke

Gruppe I
- Arado Brandenburg
- Henschel Kassel
- AEG *Im Plan seit Aug. 38*

Gruppe II
- Dornier Wismar
- Heinkel Oranienburg *Im Plan seit August 38*

Gruppe III
- Dornier Friedrichshafen *Im Plan seit August 38*

Gruppe IV
- ATG Leipzig
- Siebel Halle *Im Plan seit August 38*

Gruppe V
- Volkswagenwerk
  Fallersleben *Im Plan seit Sept. 39*

Eigene Motorenwerke
- Dessau
- Magdeburg
- Köthen
- Schönebeck
- Breslau

Lizenz Motorenwerke
- Mitteldeutsche Motorenwerke Taucha *Im Plan seit August 38*
- Pommersche Motorenwerke Stettin *Im Plan seit August 38*

Flugzeug-Rep.Werke
- Werft Leipzig *m Messehallen*
- Werft Breslau

Motoren-Reparaturwerke
- Leipziger Messe *Im Plan seit Sept. 39*

Hilfs-Werke
- Opel Rüsselsheim *Im Plan seit Sept. 39*
- Volkswagenfabrik Fallersleben *Im Plan seit Sept. 39*

Insgesamt 26 große Werke

- Hydraulikteile
- Fahrwerksteile
- Federbeine
- Leichtmetall-Gußteile
- Leichtmetall-Preßteile
- Stahl-Guß u. Preßteile
- Tm Gurte
- Propeller u. Verstellgeräte
- Zünd- u. Anlaßanlage
- Kühler Klappenantrieb
- Einspritzpumpen
- Geschützte Behälter
- Kühler
- Geräte u. Instrumente

430 Zellenteile-Unterlieferanten

- Kurbelwellen
- Zahnräder
- Guß- u. Kleingußteile
- Ventiliriegel u. Federn
- Boschteile
- Kugellager
- Lagerschalen
- Schrauben-Kleinteile
- Laufbüchsen
- Nockenwellen
- Ladedruckregler
- Öl- u. Kühlstoffpumpen

450 Motorenteile-Unterlieferanten

- Baumaterialien
- Facharbeiter
- Werkzeugmaschinen
- Käufe und Bestellungen im Protektorat
- Eisen
- Neugründungen und Ausbau von Industriewerken
- Aluminium
- Kapazitätssteigerung von Leichtmetall-Gießereien
- Finanzierungen

26

RLM	Änderungsanweisung für abgenommenes Gerät Baumuster und Baureihe **Ju 88** A-4, A-4 trop, A-8, C-6, D-1, D-1 trop	Nr. 1094 Seitenzahl: 3 Seite 1
Vorgang: Deutsches Museum	Genehmigte Änd.-Anw.v. RLM bei JFM eingegangen am: 5. November 1941 Ausgegeben am:15.11.41	Durchzuführen von: Halter Deutsches Museum	Dringlichkeitsgrad:

Nur für den Dienstgebrauch !

Betr.: Führerraum- und Tragflügel-Beheizung
Anbringung je einer Hutze über die Warmluft-Austrittsöffnungen an der linken und rechten Fahrgestellverklei-
dung

Über genannte Warmluft-Austrittsöffnungen am äußeren
Hohlkehlblech „H" der linken und rechten Fahrgestellverklei-
dung ist je eine Hutze, Teil 1, 2 und 6, 7, gemäß den Angaben
auf Seite 2 anzubringen.

Durch diese Maßnahme wird vermieden, daß bei bestimm-
ten Flugzuständen und Drehschieberstellungen Auspuffgase durch
die Luftheizung in den Führerraum und Tragflügel eindringen.

Änderungsdurchführung

Herstellerwerk	Abgenommene Werk-Nr. (nachträglich zu ändern)	Im Neubau durchgeführt	
		ab Werk-Nr.	ab Vorratssatz-Nr.
JFM Flugzeugbau	A-4, D-1 bis 0881509	0881510	
Heinkel Oranienburg	A-4 bis 0886594	0886595	
ATG Leipzig	A-4 bis 0885609	0885610	
Siebel Halle	A-4 bis 0888614	0888615	
Dornier Wismar	A-4 bis 0884599	0884600	
Henschel Berlin – Schönefeld	A-4 bis 0883624	0883625	
Arado Brandenburg	A-4 bis 0882594	0882595	
Dornier Neuaubing bei München	-	nicht gebaut	

Anmerkung: Die endgültigen Werk-Nr.-Angaben sind aus dem jeweils letzten Nachtrag
zur Übersichtsliste ersichtlich.

Unterschriften

Junkers **Flugzeug- und -Motorenwerke AG.** **Flugzeugbau** Technischer Vertrieb Kundendienst Dessau, den gez. Preitz	E'Stelle E2: gez.Kabisch 27.10.41	LC 2, II: gez.i.V.Schwerdtfeger 25.10.41 LC 2, V: gez.i.V.Rietz 25.10.41
		LC 2-Chef: gez.Reidenbach 5.11.41

World Record!
The Ju 88 as an Instrument of Propaganda

The Ju 88 was relatively late in hitting the headlines, even though it was the most modern aircraft of its day in Germany. Even though nine prototypes had been test flown by the time the Junkers propaganda film *Metallene Schwingen* appeared in 1938, the Ju 88 did not appear in the film. Instead it concentrated on the production of three aircraft types: the Ju 87, Ju 52, and Ju 86, which received the bulk of the film's attention.

The Ju 88 began making headlines on 20 March 1939. On that and following days many newspapers in Germany and abroad featured articles about the Junkers bomber that had set a world record.

Pilot Ernst Seibert and Dipl.-Ing. Kurt Heintz flew from Dessau to the Zugspitze and back at an average speed of 517 kilometers per hour, exceeding the previous record by 43 km/h. Other records followed, prompting the licensing and patent department to issue a press release on 1 August 1939. This characterized the Ju 88 as the best bomber in the world at that time, to which only the Handley Page HP 52 Hampden could be compared. This was in fact true, as the Ju 88 did not need to fear comparisons. Its contemporaries included the previously mentioned Hampden and the Vickers Wellington from Great Britain, the A-29 Hudson, B-25 Mitchell, and B-26 Marauder from the United States, the French MB-131, and the Tupolev ANT-37 from the Soviet Union. Like the Ju 88, all were twin-engined bombers.

When the campaign in the west began in 1940, the Ju 88 was repeatedly featured in war reports. In September 1940, in an article titled "It Was a Sunday Again," the *Anhalter Anzeiger* reported that the *Reichsmarschall* had personally launched an operation in which more than one million kilograms of bombs would be dropped on London. The editor summed up by saying that the attacks were in response to the despicable British raids on Dessau and other cities in the Reich. Only a few days later *Der Mitteldeutsche* of Dessau wrote that the Ju 88 had significantly enhanced the fighting strength of the *Luftwaffe*. Other German newspapers wrote about the latest bomber. In its October 1940 issue, the *Junkers Nachrichten* published an article titled *"30 Jahre Junkers-Flugzeugforschung"* in which it described in words and photographs the Ju 88's operations against England. It was especially ironic that this article appeared in both German and English.

International publications, such as *Sport Zurich* and *Fliegerwelt* of Haarlem, also wrote about this universal bomber, which could be used in a wide variety of offensive roles. In 1941 German daily and weekly newspapers wrote at length about the aircraft factory in Dessau. Titles included "A Visit to the Forge of the German Air Force," "The Birth of the Dive-Bomber," and "Ju 88 – A Victory for German Labor." In its June 1942 issue the *Kölnische Illustrierte Zeitung* featured the cockpit of the Ju 88 on its cover. Interestingly, this was the first photograph to illustrate a female company pilot, Lore Witte, from the Dessau factory.

Another popular type of article described combat pilots visiting Ju 88 production sites and thanking the workers for their efforts and high-quality product. Such articles bore titles such as "Fulfillment of Duty on the Home Front," "Successful Ju 88 Crew Visits our Factory," and "Our Aircraft and Engines in

JUNKERS

30 JAHRE JUNKERS-FLUGZEUGFORSCHUNG

JAHRGANG 31 / HEFT 10 OKTOBER 1940

NACHRICHTEN

HAUSMITTEILUNGEN DES KONZERNS DER JUNKERS FLUGZEUG- UND -MOTORENWERKE A.-G.

Cover of the Junkers company magazine from October 1940.

Clippings from European newspapers reporting on record-setting flights involving the Ju 88.

VOM EINSATZ
DER JUNKERS-JU88
GEGEN ENGLAND

JUNKERS-JU 88 ATTACKING ENGLAND

Nachtstart einer Ju 88
gegen England

Posterior take-off of a
Ju 88 against England

Foto: Albrecht

Cover of the Junkers company magazine from October 1940.

The Ju 88 V 24, first flight on 30 July 1940, photographed on the Junkers company airfield in Dessau.

Action in the German Defensive Struggle." These headlines show how the National-Socialist journalists cleverly used the Ju 88 in propaganda for the realization of their plans of conquest. Almost any means was justified in achieving this aim. One common practice was the retouching of photographs by the Junkers photographic laboratory in Dessau.

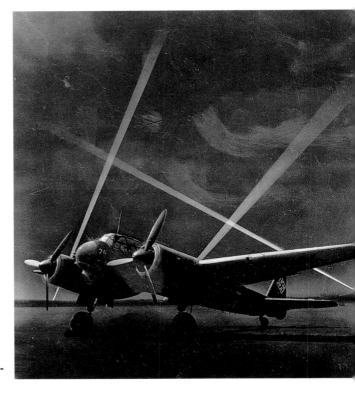

The same photograph retouched to depict a Ju 88 night fighter with search-light beams in the background.

Under Test:
The Prototypes of the Ju 88 and the
Erprobungsstelle Rechlin

The first prototype of the Ju 88, the V1, took to the air for the first time on 21 December 1936, just 12 months after the start of construction. At the controls was the veteran Junkers pilot *Flugkapitän* Karl-Heinz Kindermann. As the desired Junkers engines were not available, the aircraft was powered by two DB 600 B engines each producing 1,000 HP.

The prototype's streamlined cockpit covered a three-man cockpit, and it was capable of carrying a 500 kg payload at a speed of 450 km/h. Range was 2000 kilometers. The aircraft underwent tests at Rechlin from April to July 1937. The V1 subsequently served mainly as a test-bed and was later used in preliminary experiments for the Ju 288 program.

The second prototype of the Ju 88, the V2, was completed in April 1937. Largely similar to the V1, it was tested at Rechlin from May to August 1937.

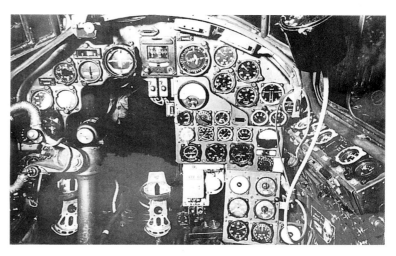

Cockpit of the Ju 88 V 1.

The third prototype, the V3, was the first to be powered by the Jumo 211, a power plant producing 1000 HP. The aircraft was equipped with typical Junkers variable-pitch propellers, large spinners, and annular radiators, which also housed the oil coolers. All this resulted in an increase in maximum speed of 16 km/h. The V3 made its first flight at Rechlin on 13 September 1937. From the beginning care was taken with this prototype to achieve the best aerodynamic form, as a result of which it was capable of an average speed of 504 km/h and even 520 km/h for short periods at a takeoff weight of 7 metric tons.

The staff at Rechlin were so satisfied with the three prototypes that Junkers Dessau planned to use the V3 for

an attempt at the world speed record with a 2000 kg payload over a distance of 1000 kilometers. 1000 kilometers was equivalent to the distance from Dessau to the Zugspitze and back. On 24 February 1938 pilot Ernst Limburger and engineer Karl-Friedrich Schonnefeld took off from Dessau and set course for the Bavarian Alps. A sudden engine failure interrupted the attempt, and the aircraft was forced to make a single-engine landing at Fürth. On touchdown the aircraft veered off course, collided with a hangar, and caught fire. The pilot and test engineer were killed in the accident. This incident highlighted the problem of a single-engine landing in the Ju 88, a procedure which demanded great skill on the part of the pilot. Similar problems were to prove fateful for numerous pilots during the Second World War.

The Ju 88 V 1 during engine test runs at Dessau, December 1936.

The Ju 88 V 1 during flight trials.

Following some design changes, on 2 February 1938 the Ju 88 V4, powered by two Jumo 211 A engines, took off on its maiden flight. It was equipped with dive brakes for flight tests and the wings incorporated larger fuel tanks for a range of 1700 kilometers. The V4 was also the first prototype to include bomb racks between the fuselage and engine nacelles. The streamlined nose cap was abandoned in favor of a revised plexiglass nose which offered better visibility. The addition of armament reduced the aircraft's maximum speed to 450 km/h. The next prototype, the Ju 88 V5, was specially designed for record-breaking flights and made its first flight on 13 April 1938. The aircraft was essentially similar to the V3. On 19 March 1939 the V5 set two world records.

Pilot Ernst Seibert and test engineer Kurt Heintz flew a distance of 1000 km, equivalent to Dessau - Zugspitze - Dessau, with a 2-ton payload at an average speed of 517 km/h. In the period that followed, additional prototypes were developed and tested at Dessau, with pilot Rupprecht Wendel carrying out most of the initial test flights. The aircraft were then flown to the *Luftwaffe* testing center at Rechlin.

The *Erprobungsstelle Rechlin* had been active as a facility of the *Reichsverbandes der deutschen Luftfahrtindustrie* (National Organization of the German Aviation Industry) since the end of the 1920s, and it was officially designated a *Luftwaffe* test center in 1935. Although it was subordinate to the *Reichsluftfahrtministerium*, the technical testing station was a "civilian operation." Rechlin was responsible for airframe testing, engine and power plant testing, aerial navigation and aviation signals testing, evaluating and checking aircraft instruments and on-board equipment, testing of aircraft guns and ammunition, and testing bombs and related aiming devices, as well as the testing of ground equipment, such as fuel trucks, airfield lighting, fire fighting equipment, and camouflage netting. The center also tested aviation fuel and lubricants.

While the prototypes were mainly tested by male pilots at Dessau, at Rechlin the situation was completely different. There a woman carried out much of the testing of the Ju 87 and Ju 88 in the dive-bomber role. She was Melitta Countess Schenk von Stauffenberg, an engineer and the second woman in Germany to be appointed *Flugkapitän*. The diving tests were usually initiated from a height of 5000 meters with a pull-out at 1000 meters in Class B and C aircraft with various types of experimental bombsights. The pilot was exposed to great physical and psychological stresses. The flights were further complicated by the fact that the aircraft were all experimental machines and technical problems were common, for example, failure of the automatic pull-out mechanism.

The Ju 88 V 1 during flight trials, here with armament installed for drag measurements.

Ju 88 A-5/trop as operated by KG 1 in the Mediterranean Theater.

Ju 88 bomber in flight.

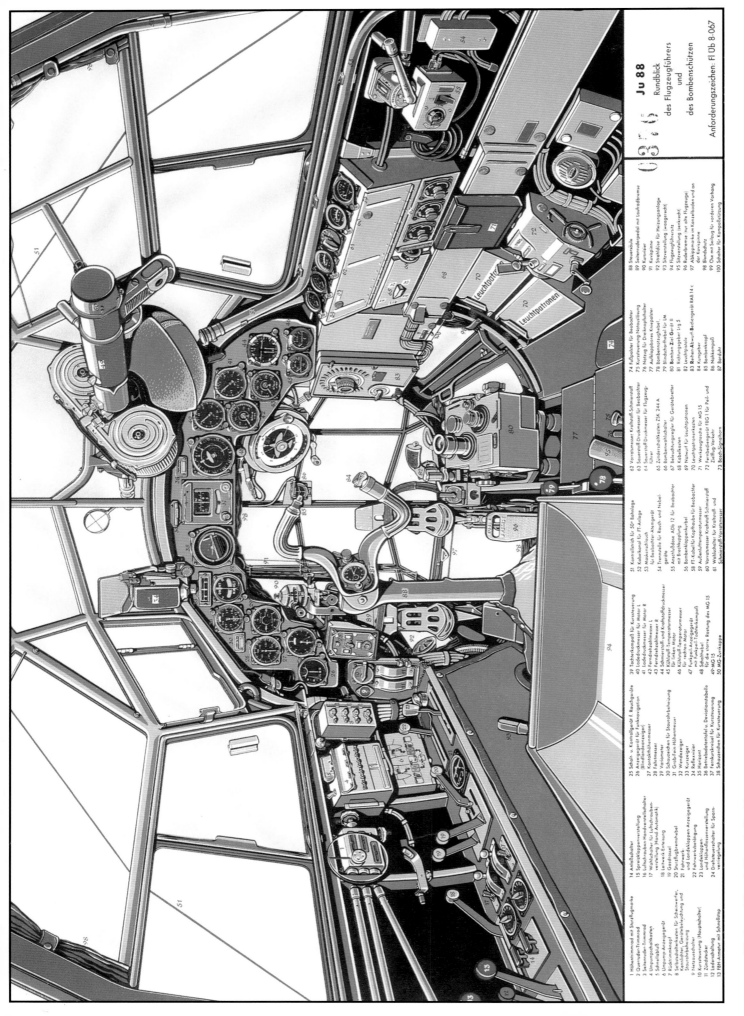

Ju 88

Rundblick
des Flugzeugführers
und
des Bombenschützen

Anforderungszeichen: Fl Ub 8-067

1 Höhentrimmrad mit Sturzflugmarke
2 Querruder-Trimmrad
3 Seitenruder-Trimmrad
4 Umpumpschalter
5 Schnellablaß
6 Leitwerk-Entlüung
7 Rückstrimklappe
8 Selbstschalterkasten für Scheinwerfer,
Kennlichter, Gerätebeleuchtung und
Staurohrbeheizung
9 Kennlichter, Geräteleuchtung und
Staurohrbeheizung
10 Kurssteuerung (Hauptschalter)
11 Zündsockel
12 Ladeschalter
13 FBH-Armatur mit Schnellstop

14 Anlaßschalter
15 Spreizklappenverstellung
16 Luftschrauben-Mandverstellung
(Blindlandeanzeiger)
17 Wahlhebel für Luftschrauben-
verstellung (Hand-Automatik)
18 Umpump-Anzeigegerät
19 Gasdrossel
20 Sturzflugbremshebel
21 Fahrwerk
22 Fahrwerk- und Landeklappen
und Landeklappenverstellung
23 Landeklappen
24 Drehzeugschalter für Spann-
regelung

25 Seitsch- u. Kontrollgerät f. Rauchgeräte
26 Anzeigegerät für Funknavigation
(Blindlandeanzeiger)
27 Kontaktöhhammesser
28 Fahrtmesser
29 Variometer
30 Schauzeichen für Staurohrbeheizung
31 Grob-Fein-Höhenmesser
32 Wendezeiger
33 Kurszeiger
34 Raffmesser
35 Horizont
36 Betriebsdatentabelle u. Deviationstabelle
37 Fernkurskreisel für Kurssteuerung
38 Schauzeichen für Kurssteuerung

39 Taschenkompaß für Kurssteuerung
40 Stilddirdukmesser für Motor L
41 Lüftdruckmesser für Motor R
42 Ferndrehzahlmesser für Flugzeugführer
43 Ferndrehzahlmesser L
44 Schmierstoff- und Kraftstoffdruckmesser R
45 Schmierstoff-Temperaturmesser
für linken Motor
46 Kühlstoff-Temperaturmesser
für rechten Motor
47 Funkgerät-Anzeigegerät
mit Funkpeil-Taschenkompaß
48 Saßaltheler
49 MG 15
50 MG-Zurückgabe

51 Kontallrich für 50r Beleilage
52 Kabankanal für FT-Anlage
53 Maskanschlauch
54 Trennanlage für Rauch- und Nebel-
gerät
55 Anschlußdose ADb 12 für Beobachter
mit Brechkupplung
56 Bombenklappenkurbel
57 Kühlstoff-Temperaturmesser
58 FT-Kabel für Kopfhörbe für Beobachter
59 Außwahlstangenanschluß
60 Vorratsmesser Kraftstoff Schmierstoff
61 Wahlschalter für Kraftstoff- und
Schmierstoff-Vorratsmesser

62 Vorratsmesser Kraftstoff-Schmierstoff
63 Saugstoff-Druckmesser für Beobachter
64 Saugstoff-Druckmesser für Flugzeugführer
65 Zünderschaltkasten ZSK 244 A
66 Bombenwahlhebel
76 Blindschalthebel
79 Blindschalthebel für LM
80 Bohrenausskäger für Gerätebretter
81 Rührungsgeber Lrg 5
82 Leuchtpistole
83 Bohren-Abwurf-Bestengerät BAB 14 c
84 Kungeber
85 Bombenknopf
86 Nahkompaß
87 Borduhr

88 Steuersäule
89 Schwenkerpedal mit Laufradbremse
90 Kurssäine
91 Kurssäine
92 Strahldüse für Heizungsanlage
93 Sitzverstellung (waagerecht)
94 Sitzverstellung (senkrecht)
95 Sitzverstellung (senkrecht)
96 Flugzeugführersitz
97 Abkippanlage am Konsatlboden und an
der Kursäine
98 Bordschutz
99 Ose und Seitzug für vorderen Vorhang
100 Schalter für Kompaßlultung

Pilot's and Bomb Aimer's Positions

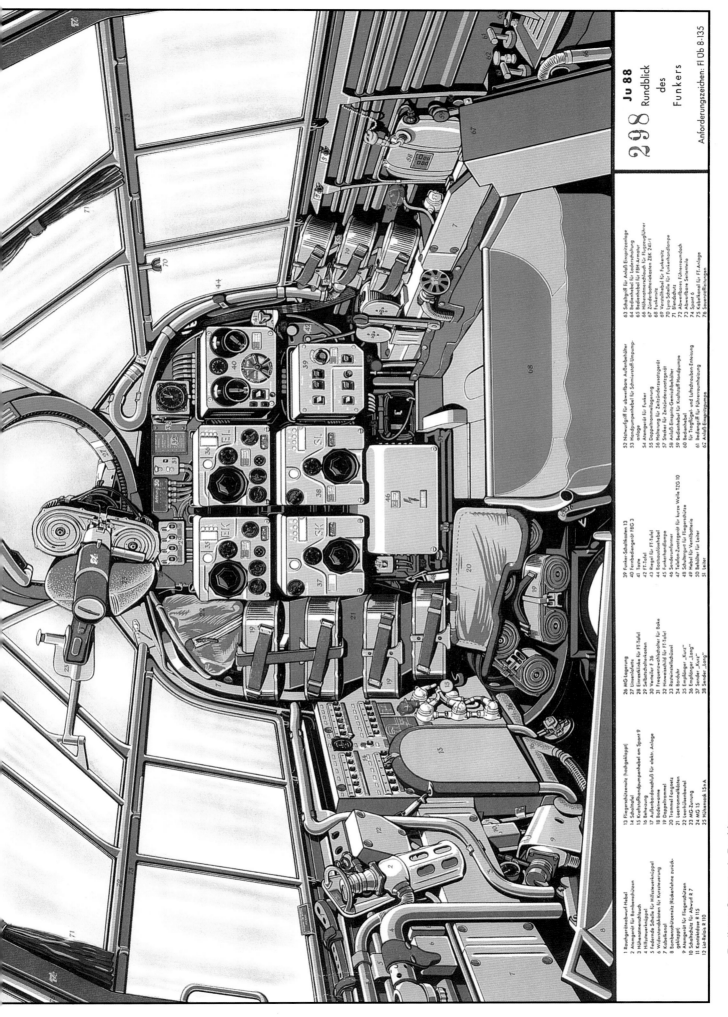

Radiator-Operator's Position

1 Rauhgerätabwurf-Hebel
2 Atemgerät für Bombenschützen
3 Höhenatemschlauch
4 Hilfssteuerknüppel
5 Federnde Scheibe für Hilfssteuerknüppel
6 Widerstandkasten für Kunststeuerung
7 Kabelkanal
8 Bombenschützensitz (Rückenlehne zurück-geklappt)
9 Atemgerät für Fliegerschützen
10 Schaltdose für Abwurf R 7
11 Kontaktdose R 110
12 Lori-Relais R 110
13 Fliegerschützensitz (hochgeklappt)
14 Schalthebel
15 Kraftstoffhandpumpenhebel am Spant 9
16 Beheizung
17 Außenbrandanschluß für elektr. Anlage
18 Außenwanne
19 Doppeltrommel
20 Trommel Fangsatz
21 Leertrommelkästen
22 Leerhülsenbeutel
23 MG-Zurrung
24 MG 15
25 Hülsensack 15nA

26 MG-Lagerung
27 Linsenlafette
28 Einrastklinke für FT-Tafel
29 Saßanschaltkasten
30 Verteiler r 30
31 Frequenzverteilschalter für Bake
32 Hinweisschild für FT-Tafel
33 Ratteinstellkästel
34 Borduhr
35 Empfänger „Kurz"
36 Empfänger „Lang"
37 Sender „Kurz"
38 Sender „Lang"
39 Funker-Schaltkasten 13
40 Fernbediengerät FBG 3
41 Taste
42 FT-Tafel
43 Riegel für FT-Tafel
44 Rastverschlußriegel
45 Funkerkabeltasche
46 Funkerhandlampe
47 Telefon-Zusatzgerät für kurze Welle TZG 10
48 Schaltergut für Fliegerschütze
49 Hebel für Ventilbatterie
50 Behälter für Leiter
51 Leiter
52 Notwurfgriff für abwerfbare Außenbehälter
53 Handpumpenhebel für Schmierstoff-Umpump-anlage
54 Atemgerät für Funker
55 Doppelstrommaßanlage
56 Halterung für Zeitzünderzusatzgerät
57 Stecker für Zeitzünderzusatzgerät
58 Anlaß-Einspritz-Gemischbehälter
59 Bedienhebel für Kraftstoff Handpumpe
60 Bedienhebel für Tragflügel- und Luftschrauben-Enteisung
61 Bedienhebel für Führerraumheizung
62 Anlaß-Einspritzpumpe

63 Schaltgriff für Anlaß-Einspritzanlage
64 Bedienhebel für Ladersschaltung
65 Bedienhebel für FBH-Armatur
66 Höhenatemschlauch für Flugzeugführer
67 Zünderbatteriekasten ZBK 241-1
68 Funkersitz
69 Verstellhebel für Funkersitz
70 Lyra-Schnelle für Funkerhandlampe
71 Blendschutz
72 Abwerfbares Führerraumdach
73 Abwerfbare Seitenteile
74 Spant 6
75 Kabelkanal für FT-Anlage
76 Sauerstoffleitungen

Ju 88 bombers on a Sicilian airfield at the base of Mount Etna.

Ground crew check the undercarriage of a Ju 88 on a Sicilian airfield.

Junkers Ju 88 on a desert airfield.

Ju 88 of a long-range reconnaissance *Staffel*.

Ju 88 long-range reconnaissance aircraft on an airfield in Sicily.

Ju 88 over the Atlantic.

Date	Time	Location	Pilot	Observer	Type/Purpose
03/10/41	1524-1620	Rechlin	Stauffenberg	Schwenk	Ju 88 code PC+EM, diving trials with four SC 250 bombs and BZA 1
10/11/41	1550-1635	Rechlin	Stauffenberg	Schwenk	Ju 88 code PC+EM, diving trials with ten SC 50 bombs and BZA 1
11/11/41	1150-1255	Rechlin	Stauffenberg	Schwenk	Ju 88 code PC+EM, diving trials with ten SC 50 bombs and BZA 1
02/12/41	1500-2880	Rechlin	Stauffenberg	Schwenk	Ju 88 code PC+EM, diving trials with four SC 250 bombs and BZA 1

It should be noted that the abbreviation SC stood for *Sprengbombe Cylindrisch* (high-explosive bomb, cylindri-cal), while the accompanying number indicated the weight of the bomb. BZA stood for *Bombenzielanlage* (bomb-aiming system), a device which underwent constant development during the war.

The difficulties associated with such flights is depicted in the following account by Dipl.-Ing. Walther Ballerstedt, one of the veterans of Rechlin:

In dive-bombing the bomb maintains its direction and speed when it is released, while the aircraft pulls out, meaning that its flight path is returned to the horizontal. When released from a vertical dive, for example from the Ju 87, the bomb continues straight down to the target. Dropped from a less steep dive, the bomb again continues in a straight line, however gravity alters its path and as a rule it falls short of the target. One must therefore allow for this when the bomb is released, making a gravity drop correction. This can be done while approaching the target by aiming over the sight, aligning the aircraft's longitudinal axis with a point beyond the target, or by aiming at the target and then pulling up the aircraft somewhat, thus reducing the gravity drop correction. The gravity drop correction is usually made at the start of the pull-out process. The gravity drop correction, or the degree to which the aircraft longitudi-

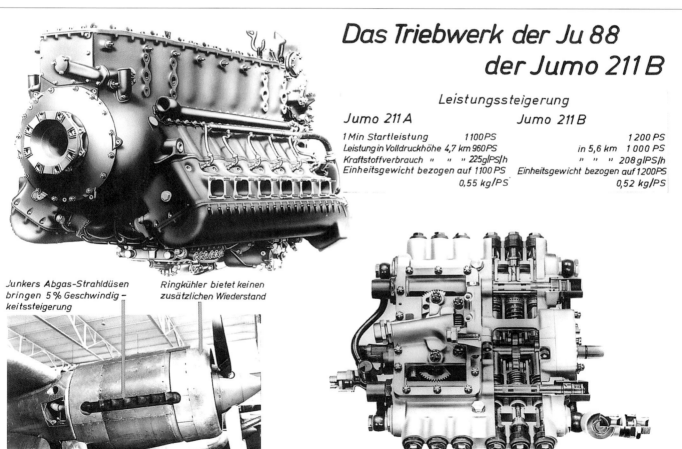

Junkers Abgas-Strahldüsen bringen 5% Geschwindig-keitssteigerung

Ringkühler bietet keinen zusätzlichen Wiederstand

Das Triebwerk der Ju 88 der Jumo 211 B

Leistungssteigerung

Jumo 211 A		Jumo 211 B	
1 Min Startleistung	1100 PS		1 200 PS
Leistung in Volldruckhöhe 4,7 km	960 PS	in 5,6 km	1 000 PS
Kraftstoffverbrauch " " "	225 g/PS/h	" " "	208 g/PS/h
Einheitsgewicht bezogen auf 1100 PS	0,55 kg/PS	Einheitsgewicht bezogen auf 1200 PS	0,52 kg/PS

Junkers Benzin-Einspritzpumpe im Schnitt

The Jumo 211 B power plant as installed in the Ju 88.

nal axis and thus the path of the aircraft and its bomb must point beyond the target, depends on the angle of dive, the release height, aircraft speed and also the headwind. The gravity drop correction angle is reached more quickly in a sharp pull-out than would be the case

in a gentle pull-out. It is therefore easy to make a mistake when dive-bombing in a fighter-bomber or other shallow dive-bomber. Through practice, however, it is possible to learn how to drop bombs in this way and to estimate with relative accuracy the gravity drop correction angle…

Erweiterte FT-Anlage und verbesserte Elt-Anlage bei Ju88B

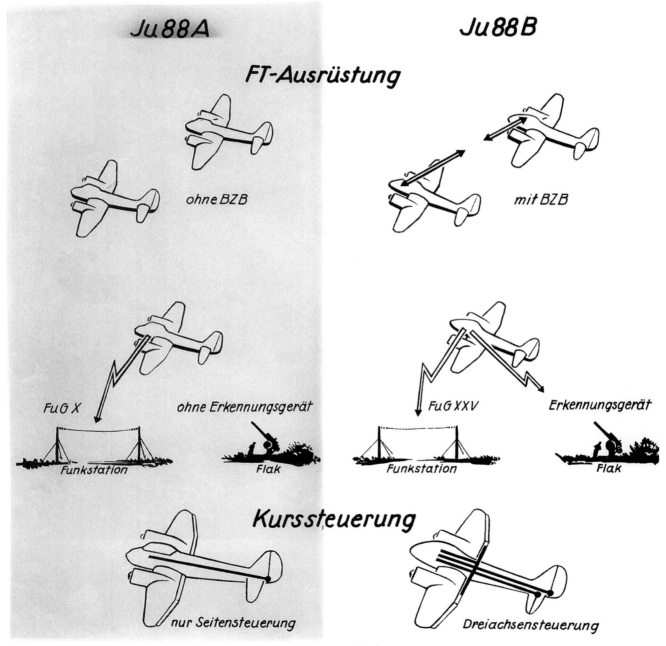

Ju88A

FT-Ausrüstung

ohne BZB

FuG X   ohne Erkennungsgerät

Funkstation   Flak

Kurssteuerung

nur Seitensteuerung

Ju88B

mit BZB

FuG XXV   Erkennungsgerät

Funkstation   Flak

Dreiachsensteuerung

Verbesserung bei Ju88B
    Erhöhung der FT-und Peilanlagen-Reichweite
    Einbau eines BZB-Gerätes (FuG XVI)
    Einbau eines Erkennungsgerätes (FuG XXV)
    Dreiachsensteuerung

Diagram illustrating the operating principle of the Ju 88's radio system.

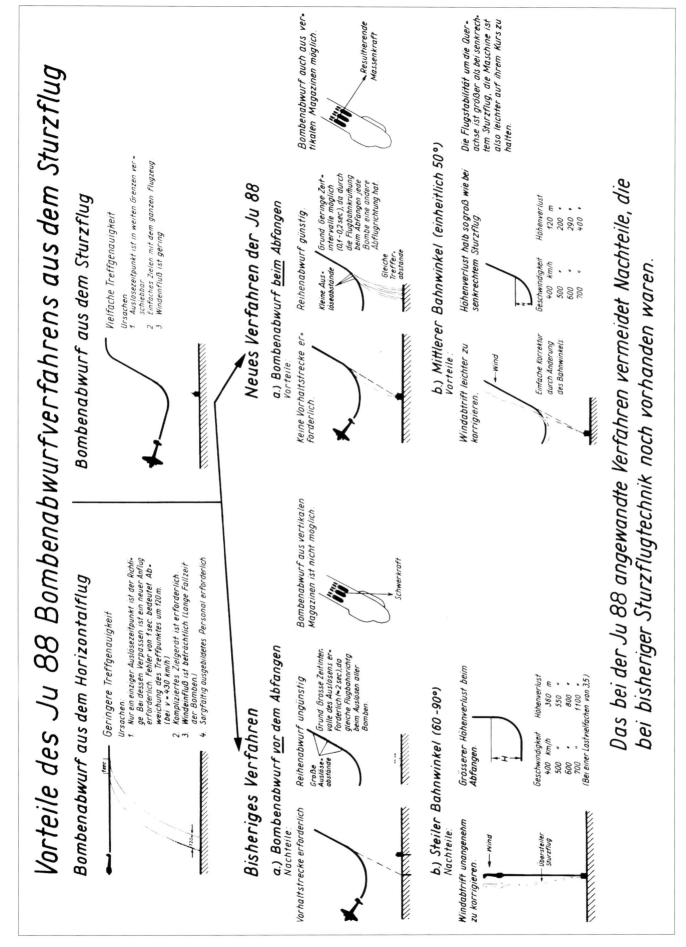

Diagram illustrating the advantages of dive-bombing compared to conventional level bombing.

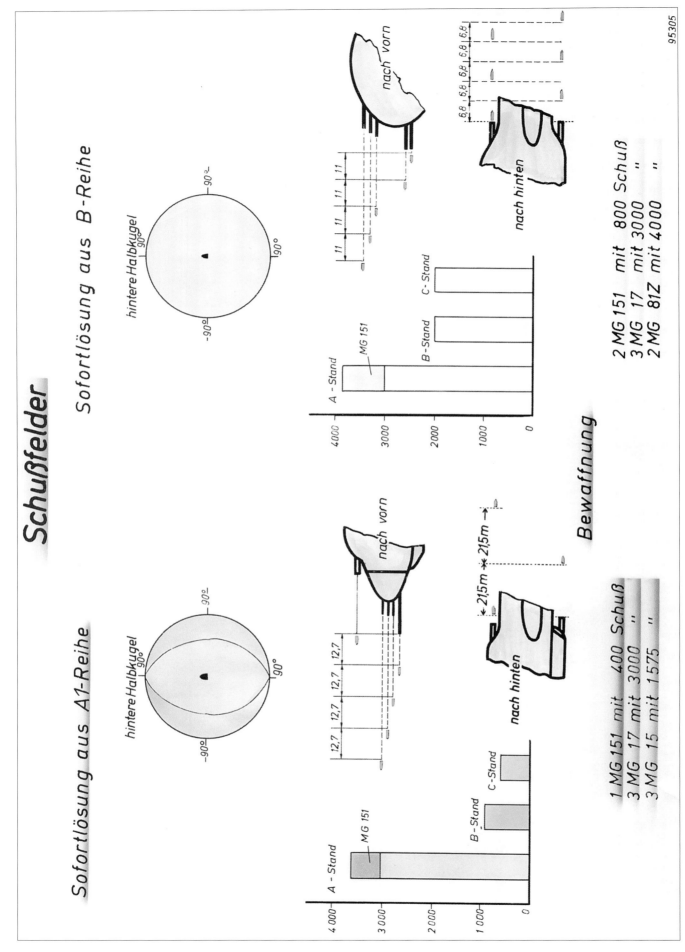

Diagram illustrating fields of fire of Ju 88 A- and B-series weaponry.

95305

When dive-bombing in the Ju 88 one had to avoid pulling up too sharply to avoid overstressing the airframe. This is why the automatic pull-out system was invented. When the bomb-release button was pressed an impulse was sent to a trim tab on the elevator. The latter moved and the aircraft pulled out at about 3 g. After a calculable time the aircraft had pulled up by a certain angle—the necessary gravity drop correction. At the same time as the impulse to the trim tab, another was sent to the bomb rack. The current did not proceed directly, instead it passed through an adjustable delay switch. In this way the impulse reached the bomb rack at the exact moment that the correct gravity drop correction angle was reached and the bomb was released at the proper time. Errors in setting the switch, in the dive angle, the release height, in estimating the headwind, etc. were always still possible…

As well as using the trim tab, it was also possible for the pilot to pull back on the control column. This would result in an excessively sharp pull-out angle, the pull-out mechanism with the release delay would miscount, and perhaps the airframe would be overstressed. To prevent this, some clever person came up with the idea of briefly disconnecting the control column when the trim tab was switched on. The pilot could pull back on the stick as hard as he wished, but the pull-out arc and bomb release process were unaffected. The control column was reconnected after a very brief interval and the pilot regained full control of the aircraft. But before it did, it is likely that even the bravest pilot received a nasty fright. Anyway, whether the system was usable or not, it had to be tested before it was issued to the units. After all, that's why Rechlin was there.

In its 6 April 1943 issue the *Luftwaffe* magazine *Der Adler* published a detailed article on the Ju 88. It was illustrated with photographs taken in Gatow in March of that year and depicted Countess Stauffenberg at the controls of a Ju 88 and in the open cockpit of a Ju 87:

> To an heroic aviatrix – the Iron Cross
> Recently Flugkapitän Dipl.-Ing. Countess Schenk von Stauffenberg was decorated with the Iron Cross and the Military Flying Badge in Gold with Diamonds. The aviatrix earned these decorations for bravery, which are seldom awarded to a woman, by risking her life and health in nearly 1,500 test flights since the start of the war. Her test flights helped in the development of air force equipment and the technical testing of German dive-bombers.

Surviving records indicate that from 1936 to 1945 115 Ju 88 prototypes were built in Dessau or were pre-assembled in Bernburg.

This Ju 88 was used as a flying test-bed for the Jumo 004 turbojet engine.

Selected Prototypes from the Ju 88 Series

Designation	Werk-Nr.	First Flight	Civil Registration	Military Registration	Power Plants
Ju 88 V1	4941	21/12/36	D-AQEN	—	DB 600 B
	Characteristics:	First prototype to be tested;			
	Later:	Testing for rapid fuel jettisoning, used in initial trials for the construction of the Ju 288 with pressurized cockpit.			
Ju 88 V2	4942	10/04/37	D-ASAZ	—	DB 600 C
	Characteristics:	Second prototype for initial trials.			
Ju 88 V3	4943	13/09/37	D-AREN	—	Jumo 211 A
	Characteristics:	Prototype built for attempt on the speed record, engine failure during record attempt on 24/02/1938. Aircraft destroyed by fire after forced landing at Nuremberg-Fürth, crew killed.			
Ju 88 V4	4944	02/02/38	D-ASYI	—	Jumo 211 A
	Characteristics:	First prototype with revised nose			
	Later:	tested with BMW 801 engines: testing of dive brakes.			
Ju 88 V5	4945	13/04/38	D-ATYU	—	Jumo 211 A
	Characteristics:	Replacement record aircraft with lengthened canopy.			
	Later:	Testing with Jumo 211 B engines: testing of dive brakes.			
Ju 88 V6	4946	18/06/38	D-AQKD	KD+ME	Jumo 211 B
	Characteristics:	Prototype for the Ju 88 A-1 version. Aircraft had a lengthened fuselage, larger horizontal stabilizer and a second bomb bay. The aircraft was used for dive-bombing trials and modifications to the cockpit and empennage. Crashed at Brandis on 13/04/44.			
Ju 88 V7	4947	27/09/38	D-ARNC	GU+AE	Jumo 211 A
	Characteristics:	Testing of airframe for later Ju 88 A-5, ETZ jettison tests, VS 11 propeller tests on 17/09/1940, preliminary trials for the Ju 88 B/E.			
Ju 88 V8	4948	03/10/38	D-ASCY	DG+BF	Jumo 211 A later Jumo 213 A
	Characteristics:	Service trials aircraft, 1940 preliminary tests for the later Ju 288, tests with balloon cable fender and balloon destroyer mechanism, aileron measurements.			
Ju 88 V9	0001	31/10/38	D-ADCN	—	Jumo 211 A
	Characteristics:	1st Zero-Series aircraft and production prototype; airframe and equipment similar to that of the Ju 88 A-1.			
Ju 88 V10	0002	03/02/39	D-AMHG	PC+CF	Jumo 211 A
	Characteristics:	2nd Zero-Series aircraft, comparison flights and performance measurements, ETZ jettisoning tests, Kutonase balloon cable cutter tests.			
Ju 88 V11	0003	2nd quarter 1939	D-AGDX	GU+AF	Jumo 211 A
	Characteristics:	Airframe and equipment similar to the Ju 88 A-1			
Ju 88 V12	0004	2nd quarter 1939		GU+AG	Jumo 211 A
	Characteristics:	First aircraft with three-axis autopilot			
Ju 88 V13	0005	12/10/39	GU+AH	—	Jumo 211 A
	Characteristics:	5th Zero-Series aircraft			

Ju 88 V14   0006   15/12/39   D-APSF   —   BMW 801 A
later BMW 801 D
Characteristics: Airframe similar to later Ju 88 A-5

Ju 88 V15   0007   1st quarter 1940   D-ARKQ   DD+IA   BMW 801 A
Characteristics: Engine test-bed with BMW in Munich, from 21/04/41 back in Dessau, converted to
Zerstörer (heavy fighter, Z-15)

Ju 88 V16   0008   1940   D-ACAR   BB+AD   Jumo 211 B
Characteristics: Tested as training aircraft, A-3 airframe, cockpit with dual controls

Ju 88 V17   0017   1940   D-AEAG   DI+MW   Jumo 211 B
Characteristics: Test-bed for high-altitude flights, tested de-icing equipment at high altitude, A-5 airframe,
testing of Kutonase

Ju 88 V18   0021   1940   D-ACAP   Jumo 213 A
Characteristics: Last aircraft of the Ju 88 series completed at Dessau, test-bed for the new Jumo 213 A
engine

Ju 88 V19   0373   1940   Jumo 211 A
Characteristics: Prototype for the *Zerstörer* version (Z-19)

Ju 88 V20   3058   1940   D-ACBN   Jumo 211 A
Characteristics: A-5 airframe for testing of flame-damper system, on 26/05/41 to Rechlin

Ju 88 V21   3113   03/11/40   D-ACBO   ND+BM   Jumo 211 J, converted to
Jumo 213 A in August 1944
Characteristics: A-4 series prototype, air and ground tests, envisaged as prototype for the Ju 88 D-1

Ju 88 V22   3132   1940   D-ACPB   ND+CF   Jumo 211 J
Characteristics: first A-4 production aircraft, later converted into Ju 188 with BMW 801 ML engines

Ju 88 V23   2001   16/06/40   D-ARYB   NK+AO   Jumo 211 B, later converted
also designated   to Jumo 213 A
V101   Characteristics: B-series prototype

Ju 88 V24   7024   30/07/40   D-ASGQ   Jumo 211 B, later converted
also designated   to BMW 801
V102   Characteristics: B-series production prototype, to Rechlin for high-altitude tests in April 1941

Ju 88 V25   7025   26/09/40   NH+AK   Jumo 211 B, later converted
also designated   to BMW 801
V103   Characteristics: B-series airframe with pressurized cockpit, to Rechlin in April 1941 for high-altitude trials

Ju 88 V26   7026   1940   NH+AL   BMW 801
also designated   Characteristics: B-3 prototype, destroyed in May 1942
V104

Ju 88 V27   7027   1941   D-AWLN   GB+NC   BMW 801, later converted
also designated   to BMW 801 ML
V105   Characteristics: E-series prototype, originally with B-series wing

Ju 88 V28   7028   1941   GB+ND   —   Jumo 213 A
also designated   Characteristics: B-series prototype with new wing and shortened cockpit
V106

Ju 88 V29   7029   1941   GB+NE   BMW 801 ML
also designated   Characteristics: prototype for the Ju 88 B-1 series, otherwise similar to the V28
V107

Ju 88 V30 also designated V108	7030	1941	GB+NF	—	Jumo 211 B
Characteristics:			high-altitude tests		
Ju 88 V31	7031	1941	GB+NG	—	Jumo 211 J
Characteristics:	prototype for equipment trials, installation of Hirth turbosupercharger				
Ju 88 V41	1004	1941	D-AFBF	DE+KD	BMW 801 MG
Characteristics:	prototype with extended wings, later converted to Ju 188				
Ju 88 V42	1042	1941	D-AFBG	DH+NE	Jumo 211 J
Characteristics:	prototype for the Ju 88 A-4/trop and Ju 88 A-11				
Ju 88 V43	1530	1941	D-AFBK	RF+JQ	BMW 801 ML
Characteristics:	prototype for the Ju 88 E-1				
Ju 88 V51	66	1942	D-ADBY	CO+OG	BMW 801 ML
Characteristics:	power plant trials, later converted to Ju 188				
Ju 88 V52	2047	1942	Jumo 211 J		
Characteristics:	prototype of the Ju 88 P-1, tests with 75-mm cannon				
Ju 88 V55	140377	1942	VL+KY	—	BMW 801 G/D
Characteristics:	prototype of the Ju 88 S-1				
Ju 88 V89	430820	1943	RG+RP	—	BMW 801 ML
Characteristics:	prototype of the Ju 88 H-1				
Ju 88 V104	710612	1943	—	—	BMW 801 D
Characteristics:	prototype of the Ju 88 G-1				
Ju 88 V107	71080	1944	VK+BM	—	DB 603 E
Characteristics:	prototype of the Ju 88 G-3				
Ju 88 V108	620151	1944	—	—	Jumo 213 A
Characteristics:	prototype of the Ju 88 G-6				
Ju 88 V112	621045	1945	—	—	Jumo 213 E
Characteristics:	prototype of the Ju 88 G-7, destroyed in air raid on Dessau on 7 March 1945				

From Reconnaissance Aircraft to Heavy Fighter: The Many Variants of the Ju 88

Altogether, 15,100 examples of the Ju 88 were built between 1939 and 1945, 7,200 of them by Junkers. The rest were manufactured under license by Arado in Brandenburg/Havel, ATG in Leipzig, Dornier in Friedrichshafen, Heinkel in Oranienburg, Henschel in Schönefeld (near Berlin), the Norddeutschen Dornier-Werke in Wismar, Siebel in Halle, and Volkswagen in Fallersleben. The aircraft's Junkers Jumo 211 engines were developed and built by the parent company in Dessau and manufactured by branch facilities in Köthen, Schönebeck, and Magdeburg, and under license by the Mitteldeutschen Motorenwerke in Delitzsch and the Pommerschen Motorenwerke in Stettin.

Junkers' branch plant in Bernburg, which in the period 1936-1937 was set up specially for the assembly of aircraft, was given the task of preparing and executing plans for quantity production of the bomber. While the parent facility at Dessau continued refining the basic design, developing new variants as the need arose and constructing prototypes, actual large-scale production of the Ju 88 began in the assembly halls in Bernburg in October 1938. The following production structure was created for Ju 88 production:

Individual parts, small components, and jigs were built in Schönebeck, Markleeberg, Prague, and Königenhof in the Protectorate of Bohemia-Moravia; Aschersleben produced the fuselages; Halberstadt delivered wings; tail assemblies were built in Leopoldshall; propellers came from Schönebeck, while final assembly of aircraft engines took place in Köthen and Magdeburg. Final assembly of the Ju

88 and acceptance flights took place in Bernburg. Only a few selected types or prototypes were built and tested by the parent facility at Dessau. A number of Junkers repair facilities were located in Schkeuditz near Leipzig and in Breslau, and as the war went on other sites were added in Germany and the occupied territories.

The first production version of the Ju 88 A-Series was the Ju 88 A-1, which entered production at Bernburg in 1939. The aircraft was powered by two Jumo 211 B engines each producing 1200 HP. The Ju 88 A-1 was conceived as a horizontal and dive-bomber with a crew of four and a bomb load of 2500 kg. Defensive armament consisted of four 7.9-mm MG 15 machine-guns.

In the years that followed, the requirements and objectives of the *Reichsluftfahrtministerium* and requests from the front-line units resulted in the appearance of numerous variants of the Ju 88 designed to fill a wide variety of roles.

The entire building process—from the manufacture of parts, to their assembly into components, larger subassemblies, and then final assembly—was subject to scrutiny so as to reduce shortcomings to a minimum. After final assembly the new aircraft was accepted by the *Bauaufsicht Luft* (BAL, or Supervision of Construction, Air) of the *Reichsluftfahrtministerium*. The acceptance process was done according to the following steps:

1. Calibrating and checking of the cockpit instrumentation, meaning the navigation equipment and compass installation, which was done on a compass base. The aircraft was placed in a level flight attitude and turned a corresponding number of degrees.

A Junkers Ju 88 A-4. Note the open access hatch with ventral gun mount and ladder.

2. After the tanks had been filled, engine tests were carried out at cruising power.
3. All problems which were discovered had to be corrected immediately.
4. The actual first flight was preceded by taxi trials, in which the aircraft was accelerated and then braked to a stop. During these the aircraft's elevator and rudder effectiveness were checked.

5. During acceptance trials all phases of flight were tested, such as climb, dive, level flight and high-altitude flight.
6. Any problems which were discovered had to be corrected immediately. Further check flights followed.

Not until all shortcomings had been eliminated did the BAL carry out the actual acceptance flight. When this had been completed the aircraft's equipment was installed and the guns test-fired. The RLM then handed the aircraft over to the receiving front-line unit.

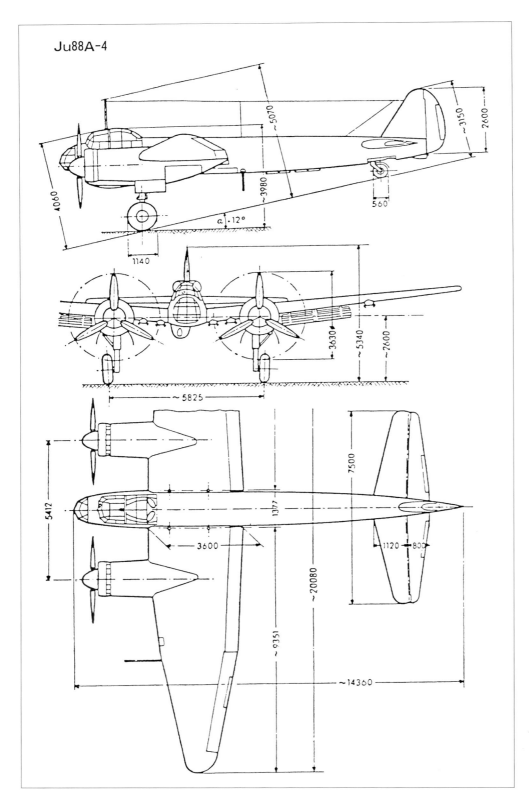

Three-view drawing of the Ju 88 A-4.

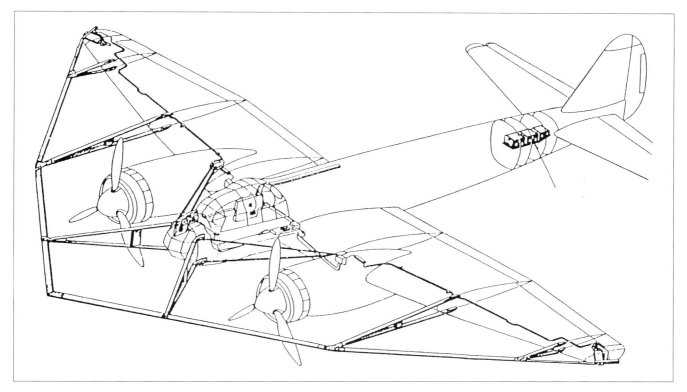

Schematic drawing of the Ju 88 A-6 with Kutonase balloon cable fender and mass balance in aft fuselage.

Ju 88 A-15 with wooden ventral bomb bay.

Front view of the Ju 88 A-15.

Führerraum Ju 88 A und Ju 88 B

Ju 88 A

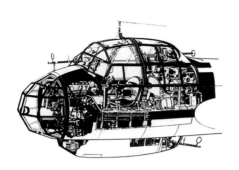

Ju 88 B

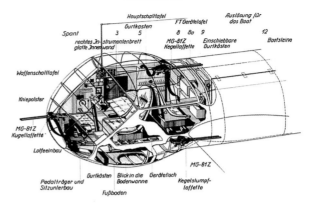

Hauptschalttafel
Gurtkasten
FT Gerätetafel
Auslösung für das Boot
Spant
rechtes Jn-strumentenbrett
glatte Jnnenwand
MG-81Z Kegellaffette
Einschiebbare Gurtkästen
Bootsleine
Waffenschalttafel
Kniepolster
MG-81Z Kugellaffette
Lotfeeinbau
Pedalträger und Sitzunterbau
Gurtkästen
Blick in die Bodenwanne
Gerätetisch
Fußboden
Kegelstumpf-laffette
MG-81Z

Führerraum Ju 88 B:

Verbesserung der Arbeitsmöglichkeit durch Vergrösserung des Raumes, glatte Jnnenverkleidung, erhöhte Bequemlichkeit. Vereinfachung der Bedienung.

Verbesserung der Sicht durch stufenlose Vollsichtglocke.

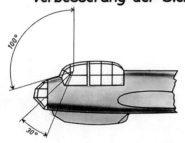

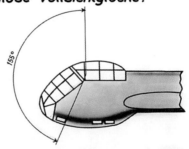

94051

Ju 88 B

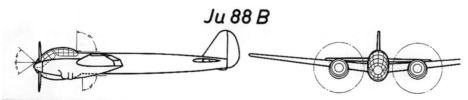

Comparison of the Ju 88 A and B cockpits. The B-series cockpit offered greater working space for the crew, increased comfort, improved ease of operation and improved view for the pilot.

KENNZEICHEN:

Neue Vollsichtkanzel
Dadurch:
Bessere Sicht
Größerer Jnnenraum
Geringerer Luftwiderstand

Stärkere Motoren
JUMO 211 F, später JUMO 213
Günstigere Triebwerkseinbauten
Neue Junkers Luftschraube VS 11
Dadurch: Höhere Geschwindigkeit
Kürzere Startstrecke

Stärkere Bewaffnung
3 Doppel-MG 81 mit Schußfolge 2800/Min
(bisher 3 MG 15 mit Schußfolge 00/Min
Munition 5000 Schuß
(bisher 2000 „)
Günstigere Schußfelder

Leistungen		JU 88 A		JU 88 B			
		mit JUMO 211 B		mit JUMO 211 F		mit JUMO 213	
Startleistung	PS	1220		1350		1500	
Kampfleistung	PS	910		1050		1240	
V max	km/h	440	460	510	525	540	555
V Reise	km/h	370	415	420	445	445	465
Reichweite	km	2900	1420	2800	1400	2800	1400
Bo	kg	1000	1400	1000	1400	1000	1400
Roll-/Startstrecke	m	685/1050		640/ 960		550/ 830	

Durch Einbau des JUMO 211 F wird der Serienanlauf der JU 88 B auf Herbst 1940 vorverlegt. Dadurch werden bis 1.4.1941 bereits 75 B Maschinen geliefert.

Chart describing the advantages of the Ju 88 B over the A-series aircraft.

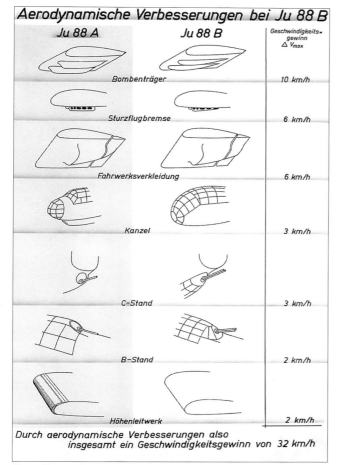

Aerodynamische Verbesserungen bei Ju 88 B

Ju 88 A	Ju 88 B	Geschwindigkeits-gewinn $\triangle V_{max}$
	Bombenträger	10 km/h
	Sturzflugbremse	6 km/h
	Fahrwerksverkleidung	6 km/h
	Kanzel	3 km/h
	C-Stand	3 km/h
	B-Stand	2 km/h
	Höhenleitwerk	2 km/h

Durch aerodynamische Verbesserungen also insgesamt ein Geschwindigkeitsgewinn von 32 km/h

Chart illustrating the aerodynamic improvements of the Ju 88 B compared to the A-series. Aerodynamic refinements listed in the chart resulted in an increase in maximum speed of 32 km/h.

Two photographs illustrating the pilot and observer positions.

Weitere Verwendungsmöglichkeiten

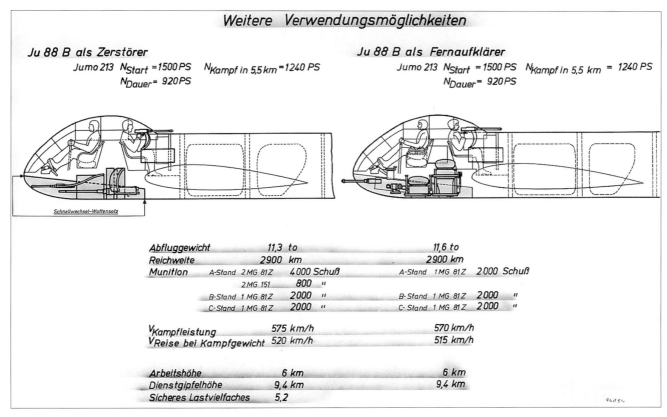

Ju 88 B als Zerstörer
Jumo 213 N_{Start} =1500 PS $N_{Kampf\ in\ 5,5\ km}$=1240 PS N_{Dauer}= 920 PS

Ju 88 B als Fernaufklärer
Jumo 213 N_{Start} = 1500 PS $N_{Kampf\ in\ 5,5\ km}$ = 1240 PS N_{Dauer}= 920 PS

Schnellwechsel-Waffensatz

		Zerstörer	Fernaufklärer
Abfluggewicht		11,3 to	11,6 to
Reichweite		2900 km	2900 km
Munition	A-Stand 2 MG 81Z	4000 Schuß	A-Stand 1 MG 81Z 2000 Schuß
	2 MG 151	800 "	
	B-Stand 1 MG 81Z	2000 "	B-Stand 1 MG 81Z 2000 "
	C-Stand 1 MG 81Z	2000 "	C-Stand 1 MG 81Z 2000 "
$V_{Kampfleistung}$		575 km/h	570 km/h
$V_{Reise\ bei\ Kampfgewicht}$		520 km/h	515 km/h
Arbeitshöhe		6 km	6 km
Dienstgipfelhöhe		9,4 km	9,4 km
Sicheres Lastvielfaches		5,2	

Proposed variants of the Ju 88 B for the heavy fighter and long-range reconnaissance roles.

Ju 88 S-1 powered by BMW 801
G radial engines.

Ju 88 G-1 night fighter equipped
with Lichtenstein C-1 radar.

Ju 88 P-2 armed with two forward-firing 37-mm cannon.

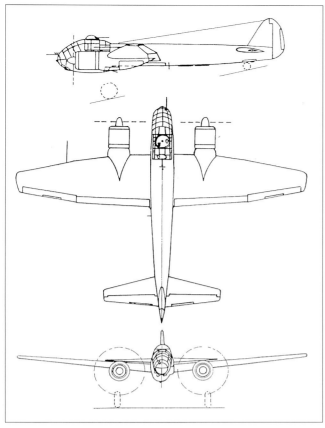

Three-view drawing of the Ju 88 B-1.

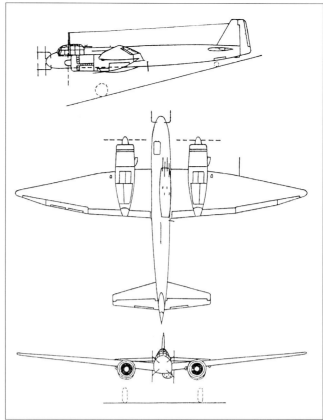

The Ju 88 G-7.

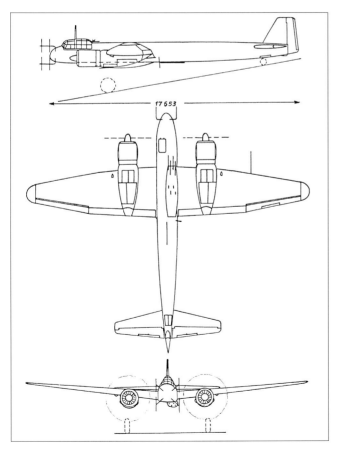

The Ju 88 H-2.

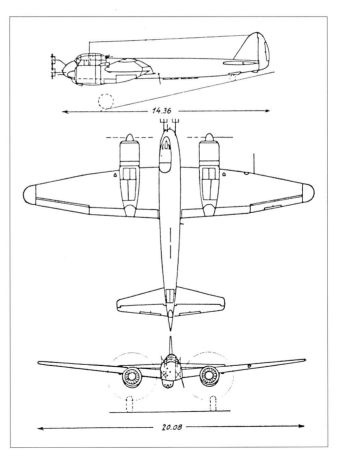

The Ju 88 R-1 and R-2.

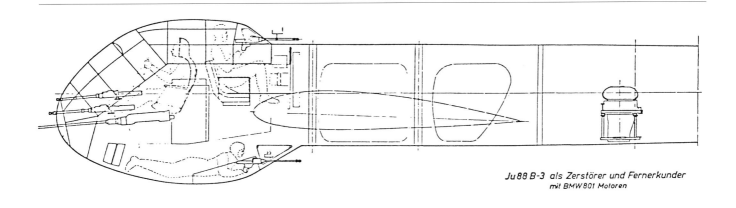

Ju88 B-3 als Zerstörer und Fernerkunder
mit BMW 801 Motoren

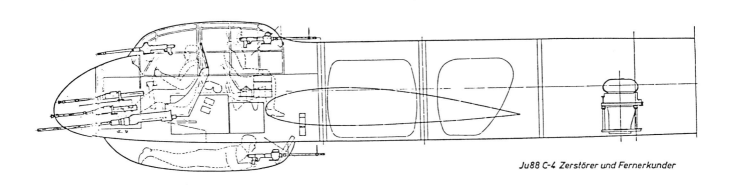

Ju88 C-4 Zerstörer und Fernerkunder

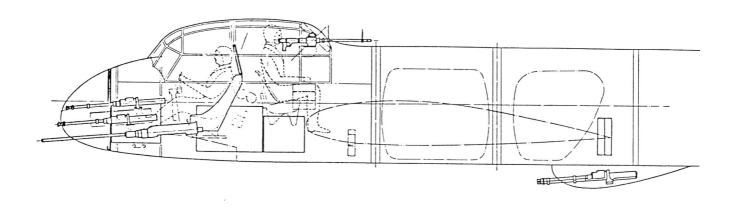

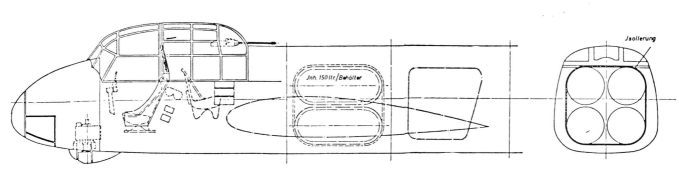

Ju 88 S-1 Schnellbomber mit BMW 801

Cockpit and armament arrangements of the Ju 88; from top to bottom: Ju 88 B-3 heavy fighter and long-range reconnaissance aircraft with BMW 801 engines; Ju 88 C-4 heavy fighter and long-range reconnaissance aircraft; Ju 88 C-5 heavy fighter; and Ju 88 S-1 high-speed bomber with BMW 801 engines.

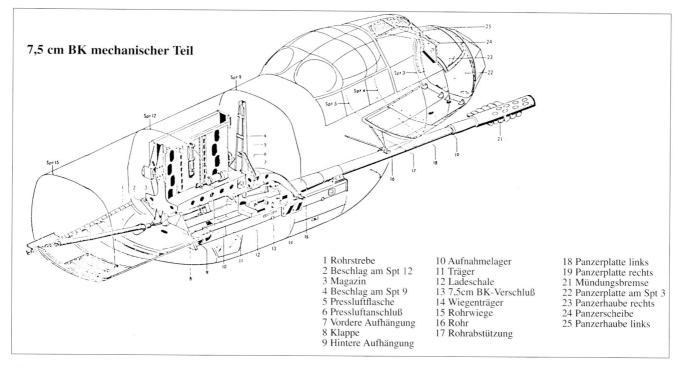

7,5 cm BK mechanischer Teil

1 Rohrstrebe	10 Aufnahmelager	18 Panzerplatte links
2 Beschlag am Spt 12	11 Träger	19 Panzerplatte rechts
3 Magazin	12 Ladeschale	21 Mündungsbremse
4 Beschlag am Spt 9	13 7,5cm BK-Verschluß	22 Panzerplatte am Spt 3
5 Pressluftflasche	14 Wiegenträger	23 Panzerhaube rechts
6 Pressluftanschluß	15 Rohrwiege	24 Panzerscheibe
7 Vordere Aufhängung	16 Rohr	25 Panzerhaube links
8 Klappe	17 Rohrabstützung	
9 Hintere Aufhängung		

Installation of the 75-mm cannon in the ventral gondola of the Ju 88 P-1.

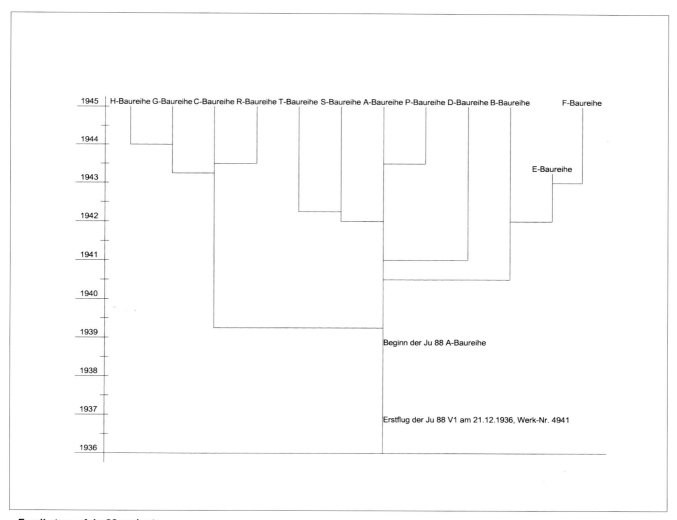

Family tree of Ju 88 variants.

Junkers Ju 88: List of Variants

Version	Role	Power Plants	Characteristics/Armament
			A-Series
Ju 88 A-0	bomber	Jumo 211 A/B	pre-production aircraft for *Luftwaffe* service trials armament: 2 x 7.9 mm MG 15
Ju 88 A-1	horizontal or dive-bomber	Jumo 211 B Takeoff power: 1,200 HP Average speed: 400 km/h at 6 000 m	Gross weight: 12 500 kg Bomb load: 2 000 kg Fuel capacity: 5 390 liters in 8 tanks, including 2 in wings, 2 in fuselage plus 2 jettisonable external tanks. Wing area: 52.5 m² Armament: 4 x 7.9-mm MG 15
Ju 88 A-2	as A-1	Jumo 211 B/G	Similar to A-1, but capable of using land catapult
Ju 88 A-3	training aircraft	Jumo 211 B/G	Training aircraft developed from Ju 88 A-1 with dual flight controls, dual throttle controls and in some cases dual instrumentation, no armament.
Ju 88 A-4	as A-1	Jumo 211 B/F Jumo 211 J rated at 1,420 HP for takeoff. Maximum speed with full bomb load: 475 km/h	Entered production in 1940, most produced later version of the Ju 88. Design constantly revised until 1944 Wing area: 54.7 m² Armament: 1 x 13-mm machine-gun in nose (A-Stand) 2 x 7.9 mm MG 81 in rear of cockpit (B-Stand) 2 x 7.9-mm machine-guns in rear of ventral gondola (C-Stand)
Ju 88 A-4 trop	as A-1	Jumo 211 J	Similar to A-4 but with special tropical equipment, such as extra water tanks, sun and mosquito protection.
Ju 88 A-5	as A-1	Jumo 211 B/D/G/H Average speed: 450 km/h	Developed from the A-1 version, with increased wing area of 54.5 m². Gross weight: 12 200 kg Bomb load: 2 400 kg Armament: 1 x 13-mm MG 131 in nose (A-Stand) 2 x 7.9 mm MG 15 in rear of cockpit (B-Stand) 1 x 13-mm MG 131 in rear of ventral gondola (C-Stand)
Ju 88 A-6	as A-1	Jumo 211 B/G	Similar to A-5 version, but with balloon cable deflector and cutter in leading edge of wing. Deflector and associated trim weights resulted in loss of speed of approximately 30 km/h. Gross weight: 13 000 kg
Ju 88 A-6/U	maritime aircraft	Jumo 211 H	Version of the A-5 for the maritime reconnaissance reconnaissance role. FuG 200 Hohentwiel search radar installed in ventral gondola. Gross weight: 11 000 kg Armament: no bomb load 1 x 7.9-mm MG 81 in A-Stand 2 x 7.9-mm MG 81 in B-Stand
Ju 88 A-7	training aircraft	Jumo 211 B/G Maximum speed: 475 km/h	Converted A-5 for the training role, equipped with dual controls. Armament: no bomb load 1 x 7.9-mm MG 15 in A-Stand 2 x 7.9-mm MG 15 in B-Stand 1 x 7.9-mm MG 15 in C-Stand
Ju 88 A-8	as A-1	Jumo 211 H	Similar to A-4 version, but with balloon cable deflector and cutter like A-6. Later converted to conventional bomber with Jumo 211 J engines.
Ju 88 A-9	as A-1	Jumo 211 B Average speed: 455 km/h	Similar to A-1 version, but with tropical equipment, such as additional water tanks, sun protection and special equipment. Gross weight: 12 300 kg, equipment weight: Tropical equipment: 190 kg Bomb load: 2 700 kg Armament: 6 x 7.9-mm MG 15

Ju 88 A-10	as A-1	Jumo 211 210 B/G	Similar to A-5 version, but with tropical equipment as A-9. Gross weight: 12 430 kg. Armament: 1 x 7.9-mm MG 15 in A-Stand / 2 x 7.9-mm MG 15 in B-Stand / 2 x 7.9-mm MG 15 in C-Stand
Ju 88 A-11	as A-1	Jumo 211 G	Equivalent to tropical version of the A-4, but with different engines. Weight: 12 100 kg with 2 200-kg bomb load.
Ju 88 A-12	training aircraft	Jumo 211 B/G	A-version converted for pilot training role. Larger cockpit with dual controls. No armament, dive brakes deleted.
Ju 88 A-13	close-support aircraft	Jumo 211 H	Developed from A-1 with increased armor and heavier armament, designed specially for low-altitude attacks. Dive brakes deleted. Gross weight: 13 000 kg, including 500 kg of SD-2 anti-personnel bombs. Armament: as many as 16 x 7.9-mm MG 17 fixed forward- or rearward-firing machine-guns, so-called "Watering Can" for strafing enemy troops.
Ju 88 A-14	horizontal bomber	Jumo 211 J	Special version of the A-4 with extended wings and balloon cable cutter (Kuto-Nase). Gross weight: 12 100 kg / Service ceiling: 8 000 m Range: 2 500 km / Armament: 2 x 7.9-mm MG 15 / 2 x 13-mm MG 131 / Some aircraft equipped with 20-mm MG-FF cannon in nose for engaging shipping targets.
Ju 88 A-15	as A-1	Jumo 211 G later Jumo 211 J	Special version of the A-14 with bulged wooden ventral fairing permitting internal loads of up to 3 300 kg to be carried.
Ju 88 A-16	training Aircraft	Jumo 211 J	Conversion of A-14 for training role with dual controls.
Ju 88 A-17	torpedo bomber	Jumo 211 J	Special version of the A-4 with additional armament of two 1 100-kg torpedoes beneath the wings. Gross weight: 11 500 kg. Armament: 1 x 7.9-mm MG 81 in A-Stand / 2 x 7.9-mm MG 81 in B-Stand / 1 x 7.9-mm MG 81 in C-Stand

B-Series

Ju 88 B-0	experimental bomber (built in small numbers only)	Jumo 211 F or Jumo 213	Development of the A-5 with streamlined, fully-glazed nose. Gross weight: 12 600 kg. Range: 2 800 km / Service ceiling: 9 000 m / Armament: 2 x 7.9-mm MG 81 in A-Stand / 2 x 7.9-mm MG 81 in B-Stand / 1 x 7.9-mm MG 81 in C-Stand
Ju 88 B-1	mock-up only		Similar to A-0 with increased armor in ventral gondola and improved defensive armament with DL 131 powered turret in B-Stand.

C-Series (also designated Z-Series)

Ju 88 C-1	heavy fighter	Jumo 211 B	Derived from A-1, but with bomb-dropping and dive-bombing systems deleted. Armament: 1 x MG 151 (fixed in nose), 3 x MG 17
Ju 88 C-2	heavy fighter and night fighter	Jumo 211 G	Based on A-1 version but with increased wingspan. Armament: 1 x 20-mm MG-FF and 3 x 7.9-mm MG 15 (fixed in nose) / 1 x 7.9-mm MG 15 in B-Stand / 1 x 7.9-mm MG 15 in C-Stand / The night fighter version also had two MG-FF cannon in ventral gondola.
Ju 88 C-3	heavy fighter and night fighter	BMW 801 A	Similar to C-2 with BMW radial engines

Ju 88 C-4	heavy fighter	Jumo 211 B	Similar to C-2 version but with installation of reconnaissance cameras (Rb 50/30, 20/30 or similar equipment) between Bulkheads 17 and 18. Installation of instrument landing system.
Ju 88 C-5	heavy fighter	BMW 801	Similar to C-2 but without ventral gun position, crew reduced to two. Armament: C-Stand replaced by two fixed forward-firing MG 17.
Ju 88 C-6	heavy fighter, day and night fighter	Jumo 211 J and BMW 801 later the FuG 202 and FuG 220	Aircraft built in different versions to suit operational requirements. Radar equipment: FuG 202 Lichtenstein or FuG 212 C1 Lichtenstein SN 2, FuG 227 Flensburg (reduced aircraft speed by ca. 40 km/h). Additional armament: 1 x 13-mm MG 131 in B-Stand 2 x 20-mm MG 151 cannon mounted obliquely in fuselage aft of cockpit, so-called "Schräge Musik."
Ju 88 C-7	heavy fighter and night fighter	Jumo 211 J	Similar to C-6 version, but with extended wings.

As the comparison of available photographs reveals, C-Series aircraft, especially the C-6 variant, were subject to numerous modifications during front-line service. This has, understandably, led to conflicting information in some publications.

D-Series

Ju 88 D-1	long-range reconnaissance	Jumo 211 J	Similar to the A-4, but with dive brakes removed and additional fuel tanks in and under the fuselage. As a result, range increased to 5 000 km. Several cameras located behind Bulkhead 15.
Ju 88 D-2	long-range reconnaissance	Jumo 211 B Jumo 211 G	Developed from A-5 Equipment similar to D-1
Ju 88 D-3	long-range reconnaissance	Jumo 211 J	Tropical version of D-1
Ju 88 D-4	long-range reconnaissance	Jumo 211 J	Tropical version of D-2
Ju 88 D-5	long-range reconnaissance	Jumo 211 J	Similar to D-1, but with VDM propellers
Ju 88 D-6	long-range reconnaissance	BMW 801 or BMW 801 J	Otherwise similar to D-2

E-Series

Ju 88 E-0	experimental bomber	BMW 801 D later BMW 801 M2	Airframe similar to A-4, but with the streamlined fully-glazed cockpit of the Ju 88 B-0.
Ju 88 E-1	bomber	Jumo 211 J	Armament: 1 x 20-mm MG 151 cannon in A-Stand 1 x DL 131 in B-Stand (fixed) 2 x 7.9-mm MG 81 in B-Stand 2 Increased engine power resulted in increased bomb load.
Ju 88 E-2	bomber	BMW 801 D	Similar to E-1

F-Series

Ju 88 F-1	long-range reconnaissance	Jumo 211 J	Similar to E-1, but without bomb racks. Camera equipment similar to D-1.

G-Series

Ju 88 G-0	prototype for the night fighter	BMW 801 D	Fuselage and tail of the Ju 188, wings taken from the Ju 88 A-4. Expanded electronics series with FuG 220 Lichtenstein SN 2. Armament: 6 x 20-mm MG 151 cannon, fixed in nose and fuselage tray.

Ju 88 G-1	night fighter	BMW 801 G	Production version of the G-0 with addition of warm air heating for horizontal stabilizer, Patin autopilot and additional fuel tank in forward bomb bay. Armament: 2 x 20-mm MG 151 cannon (fixed) in fuselage nose 4 x 20-mm MG 151 cannon (fixed) in fuselage tray 1 x DL 131 in B-Stand
Ju 88 G-2	night fighter prototype	Jumo 213 A/E	Similar to G-1 with improved radar equipment and armament (30-mm cannon). No production, because of Ju 388 development
Ju 88 G-3	night fighter	DB 603	Modified G-1, no production
Ju 88 G-4	night fighter	Jumo 213 A	Modified G-1, no production
Ju 88 G-5	night fighter	Jumo 213 A	Modified G-1, no production
Ju 88 G-6	night fighter (1,750 HP)	Jumo 213 E	Similar to G-1 with more powerful engines
Ju 88 G-7	night fighter	Jumo 213 A	Similar to G-6, but with wing of the Ju 188 and additional armament. 2 x MK 108 in fuselage tray
Ju 88 G-8	night fighter	Jumo 213 E	Modified G-1, no production
Ju 88 G-9	night fighter	BMW 801 G	Modified G-1, no production
Ju 88 G-10	heavy fighter and night fighter	Jumo 213 A	A-4 airframe with Ju 188 wings, used in Mistel combination

H-Series

Ju 88 H-1	long-range reconnaissance	BMW 801 D	A-4 airframe, extended to 17.75 meters with a larger ventral fairing resulting in a continuous bomb bay. Additional fuel tanks made possible an increase in tactical operations range to 4 800 km or 12 hours flying time. Three-axis autopilot improved handling and accuracy of weapons. Equipped like D-1. Armament: 2 x MG 131 in ventral tray 1 x MG 81 in A-Stand 1 x 13-mm machine-gun in B-Stand
Ju 88 H-2	long-range reconnaissance long-range bomber	BMW 801 G	Similar design to the G-2 version, however airframe consisted of components of the A-4 and Ju 188, resulting in length of 17.88 m with continuous bomb bay. Wingspan increased to 19.95 m.
Ju 88 H-3	long-range reconnaissance long-range bomber	BMW 801 D/G	Similar to H-2, however fuselage further lengthened.
Ju 88 H-4	long-range fighter	Jumo 213 A	Similar to H-2, but with greatly lengthened fuselage. Used in Mistel combination.

P-Series

Ju 88 P-1	tank-destroyer	Jumo 211 J	Airframe of the A-4, with bomb-dropping and dive-bombing systems removed. Large ventral fairing housing a 75-mm Pak 40 anti-tank gun. Not produced.
Ju 88 P-2	tank-destroyer	Jumo 211 J	Similar to P-1 with strengthened airframe. Armament: 2 x 37-mm Flak 38 in ventral fairing 1 x 7.9-mm MG 81Z in rear of ventral fairing
Ju 88 P-3	tank-destroyer	Jumo 211 J	Similar to P-2 with heavier armor
Ju 88 P-4	tank-destroyer	Jumo 211 P	Similar to P-3. Armament: 1 x 50-mm KwK 39

R-Series

Ju 88 R-1	night fighter	BMW 801 MA	Similar to C-6 version with more powerful engines and improved FuG 212 Lichtenstein C1 radar
Ju 88 R-2	night fighter	BMW 801 G2 or	Similar to C-6 version with more powerful BMW 801 D2 engines and FuG 220 Lichtenstein C2 radar.

Ju 88 S-1	high-speed bomber	BMW 801 D/G	Aerodynamically-refined A-4 airframe, no ventral gondola, automatic pullout system, modified glazing with spherical nose cap. Speed 600 km/h at height of 8 000 m. Additional fuel tanks, reduced armor and defensive armament of 1 x MG 131 in B-Stand. Bomb bay for 18 x 50 kg bombs.
Ju 88 S-2	high-speed bomber	BMW 801 G	Similar to S-1 with more powerful engines (two-speed supercharger with GM-1 injection). Speed 615 km/h at height of 8 000 to 10 000 meters.
Ju 88 S-4	high-speed bomber	Jumo 213 A1	Modified S-3 with components of the Ju 188. Able to accommodate larger-caliber bombs within fuselage, range and speed reduced.
Ju 88 S-5	high-speed bomber	Jumo 213 T	Similar to S-2 version. Improved engines with turbosuperchargers made possible a speed of 600 km/h at a height of 11 000 meters. Not produced.

T-Series

Ju 88 T-1	high-speed reconnaissance	BMW 801 D	Derived from the S-1, with reconnaissance equipment similar to that of the D-1.
Ju 88 T-2	high-speed reconnaissance	BMW 801 J	Similar to the T-1 with heavier defensive armament: 2 x MG 15 in B-Stand or 2 x MG 81 in B-Stand Not produced.
Ju 88 T-3	high-speed reconnaissance	Jumo 213 A or Jumo 213 E	Similar to the T-1 with more powerful engines, resulting in max. speed of 640 km/h at 10 000 m. Reconnaissance equipment consisted of: 1 x Rb 20/30 camera 1 x Rb 540/30 camera 1 x Rb 70/30 camera

The Ju 88 in Operational Service

The Second World War, which was to have such fateful consequences for Germany, began on 1 September 1939. The results achieved by the *Luftwaffe* in the opening days of fighting confirmed the importance of this arm in the military confrontation.

The Ju 87 dive-bomber achieved great success at the start of the campaign against Poland. This was in keeping with the *Luftwaffe*'s doctrine of pinpoint tactical attacks, which was in part based on experience gained in the Spanish Civil War. When the western campaign began in 1940, these tactics were still viable; however, when Germany was forced into a wide-ranging strategic air war it soon became obvious that they were outdated. The elements in the *Luftwaffe* operations staff which had been pushing for a twin-engined or larger dive-bomber since 1938 now won through. Immediately after assuming the post of Head of the Technical Office in June 1936, *Oberst* Ernst Udet demanded the standardization of airframes and engines. This was to be taken into consideration during development and production so as to make possible the exchange of parts during repairs. Parallel type development was to be excluded from the outset.

This interesting concept was doomed to failure by the rivalries that existed within the German aviation industry. Furthermore *Reichsmarschall* Göring, Commander-in-Chief of the *Luftwaffe* and the *Reichsluftfahrtministerium*, and thus Udet's superior, did not agree. In 1937, no doubt influenced in part by the Condor Legion's experiences in Spain, a requirement was issued for an increase in production of existing types, such as the Heinkel He 111, Junkers Ju 86, Dornier Do 17, and Junkers Ju 87, with the objective of reaching the following production figures by April 1938:

He 111	increase from 831 aircraft to 850
Ju 86	production unchanged at 680 aircraft
Do 17	from previous 788 aircraft to 1,014
Ju 87	from previous 264 aircraft to 345

The previously-mentioned difficulties in obtaining raw materials, such as iron, steel, aluminum, and plexiglass affected these goals to some degree. The secret conferences held by the *Generalluftzeugmeister* revealed that there was a recurring problem in planning and materials procurement which continued into the Second World War and which, in spite of harmonization and the annexing of entire branches of industry in Europe, was to grow worse by the end of the war. One potential solution was the appointment of a "Special Plenipotentiary" to overcome bottleneck situations and address material and organizational problems.

On 30 September 1938 the General Director of Junkers, Dr. Heinrich Koppenberg, received a written directive from the Commander-in-Chief of the *Luftwaffe*, *Reichs-*

Fixed forward-firing machine-gun (*A-Stand*).

Installation of the Jumo 211 in the wing of the Ju 88.

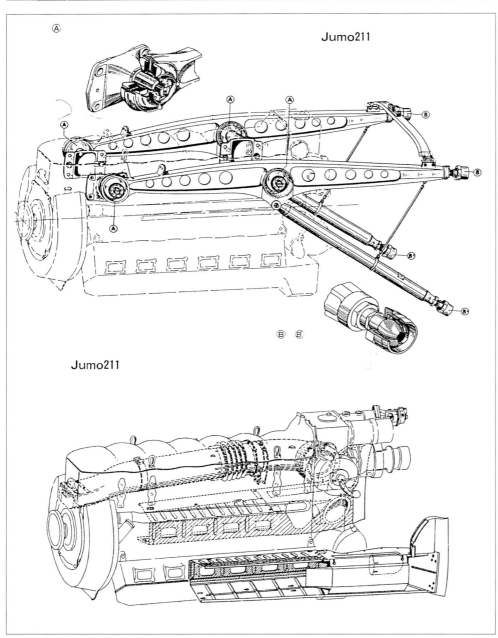

Jumo211

Jumo211

Jumo 211 engine mounts and fittings.

Twin machine-gun positions in the rear cockpit of the Ju 88 (*B-Stand*).

Armored lens mounts in the rear of the Ju 88 cockpit.

marschall Hermann Göring, instructing him to take whatever measures necessary to assure rapid production of the Ju 88. The letter made it clear that Koppenberg was now a plenipotentiary with the authority to issue directives to other firms, including subcontractors, to coordinate production of the Ju 88. This was again stressed in a letter from Reich Chancellor Adolf Hitler to Hermann Göring on 21 August 1939.

These actions gave the Ju 88 program priority over other aircraft programs. The directive was preceded by a display at Rechlin on 3 July 1939, when a number of weapons and aircraft were displayed to Hitler. He was extremely impressed by the Ju 88's potential to operate at previously unattainable altitudes. That the Ju 88's pressurized cockpit was still in the experimental stage at that point was given little importance at the display, however, that fact and other technical difficulties were to set back production of the Ju 88 by months. The published delivery program was permanently changed, and as a result the requested numbers of Ju 88s constantly had to be revised downward. For example, the modified Delivery Program No. 12 called for 2,357 examples of the Ju 88 to be built by April 1941; by April 1943 the total number was to climb to 5,000. Göring did not agree with these targets, and on 5 August 1939 he paid a surprise visit to the Junkers factories in Dessau. There he convinced the firm's leaders and employees that production of approximately 300 Ju 88s per month was altogether possible and necessary.

Göring's demand was understandable when one considers the bomber force available to the *Luftwaffe* when war broke out. The principle types were the

Interior view of the armored lens mount.

Ventral machine-gun (*C-Stand*).

Comparison of fields of fire of defensive weapons mounted in the Ju 88 A and Ju 88 B.

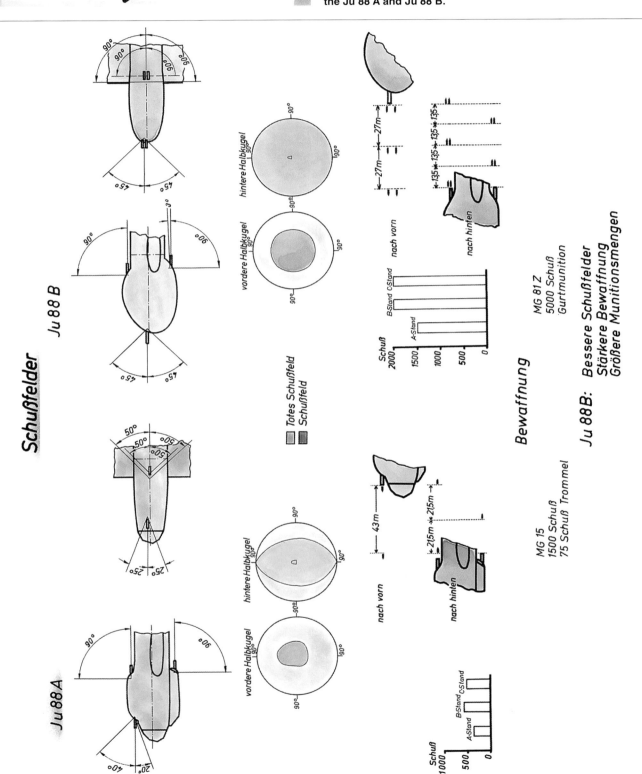

58

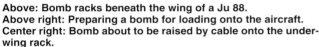

Above: Bomb racks beneath the wing of a Ju 88.
Above right: Preparing a bomb for loading onto the aircraft.
Center right: Bomb about to be raised by cable onto the under-wing rack.
Bottom right: Bomb in place on underwing rack.

Dornier Do 17 E and M, the Heinkel He 111 B, E, F, and J, and the Junkers Ju 86 A, D, E, and G, a total of 1,235 aircraft. By that time only a handful of Ju 88s had reached the combat units.

The Ju 88 saw its first major action in the western campaign against the Netherlands, Belgium, and France in May 1940. Ju 88s operated from captured airfields against targets in the Dunkirk area in an effort to interrupt the British "Operation Dynamo." It was there that the Ju 88 received its baptism of fire. The type's technical and tactical characteristics soon made it the standard German medium bomber. The main production version of the Ju 88, the A-4, entered production in the summer of 1940. The type was powered by the Jumo 211 B producing 1200 HP, and later the Jumo 211 F producing 1340 HP. Based on front-line experience it introduced a strengthened undercarriage and an increase in wingspan from 18.37 to 20.08 meters, resulting in a wing area of 54.7 m². Defensive armament was also bolstered to one 13-mm MG 131 in the nose for the observer and bomb aimer (*A-Stand*), two 7.9-mm MG 81 machine-guns in lens mounts in the dorsal position (*B-Stand*), and a twin-barreled 7.9-mm MG 81 Z in the ventral gondola (*C-Stand*). The ventral gondola was armored to protect the gunner against fire from below. As bomb loads rose steadily, fittings were installed on the undersides of the wings for jettisonable takeoff-assist rockets.

Contact between the pilots and maintenance personnel and the designers was vital for the further development of the Ju 88. This made it possible to incorporate suggested improvements quickly and constantly improve the aircraft's operational effectiveness. This sort of teamwork was an effective propaganda tool, illustrating how "an inner and an outer front" were supposed to cooperate in the spirit of National-Socialism.

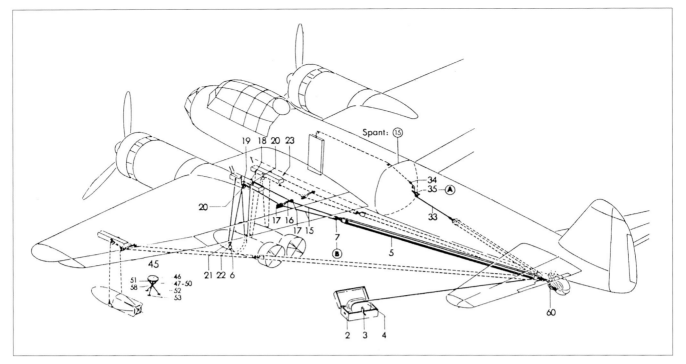

Schematic drawing of Ju 88 bomb racks.

A Ju 88 on a mission high above the clouds.

The Ju 88 A-4 was the most-produced version of the aircraft and was the standard production horizontal- and dive-bomber, slowly becoming the backbone of the German bomber force. German propaganda exaggerated the effectiveness of the *Luftwaffe* and of course that of the Ju 88. This was in part responsible for the type's reputation as a "super bomber." In fact, the first Ju 88s were only capable of penetrating to a depth of about 880 kilometers. Not until later, with the addition of larger fuel tanks, was it possible to increase their range to 1660 kilometers. The range of the Ju 88 A-1 was approximately 2500 kilometers, and the A-4 was the first version to be capable of 3150 kilometers. These gaps between actual performance and propaganda claims may still be found in the existing literature and have contributed to a false assessment of the Ju 88.

Seen from the observer's position of a trailing aircraft, a Ju 88 produces condensation trails at high altitude.

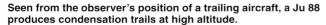

Serienausführung Ju 88 A und Ju 88 B

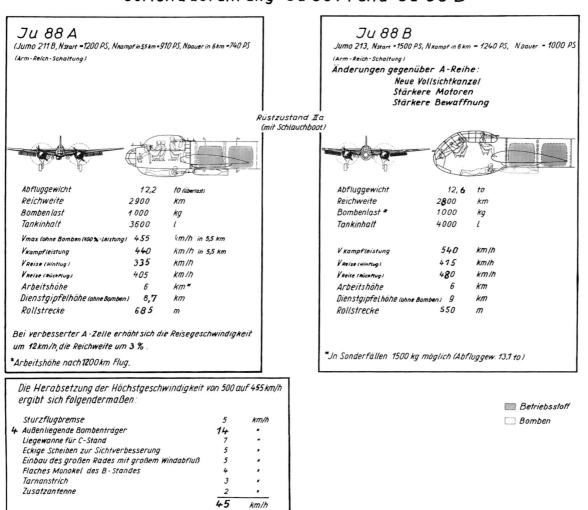

Ju 88 A
(Jumo 211 B, N_{start} =1200 PS, N_{kampf in 5,5 km} =910 PS, N_{Dauer in 6 km} =740 PS
(Arm-Reich-Schaltung)

Ju 88 B
Jumo 213, N_{start} =1500 PS, N_{kampf in 6 km} = 1240 PS, N_{Dauer} = 1000 PS
(Arm-Reich-Schaltung)

Änderungen gegenüber A-Reihe:
Neue Vollsichtkanzel
Stärkere Motoren
Stärkere Bewaffnung

Rüstzustand IIa
(mit Schlauchboot)

Ju 88 A		
Abfluggewicht	12,2	to (überlast)
Reichweite	2900	km
Bombenlast	1000	kg
Tankinhalt	3600	l
Vmax (ohne Bomben (100%-Leistung)	455	km/h in 5,5 km
Vkampfleistung	440	km/h in 5,5 km
Vreise (Hinflug)	335	km/h
Vreise (Rückflug)	405	km/h
Arbeitshöhe	6	km*
Dienstgipfelhöhe (ohne Bomben)	8,7	km
Rollstrecke	685	m

Bei verbesserter A-Zelle erhöht sich die Reisegeschwindigkeit um 12 km/h, die Reichweite um 3 %.
*Arbeitshöhe nach 1200 km Flug.

Ju 88 B		
Abfluggewicht	12,6	to
Reichweite	2800	km
Bombenlast *	1000	kg
Tankinhalt	4000	l
V kampfleistung	540	km/h
V reise (Hinflug)	415	km/h
V reise (Rückflug)	480	km/h
Arbeitshöhe	6	km
Dienstgipfelhöhe (ohne Bomben)	9	km
Rollstrecke	550	m

*In Sonderfällen 1500 kg möglich (Abfluggew. 13,1 to)

Die Herabsetzung der Höchstgeschwindigkeit von 500 auf 455 km/h ergibt sich folgendermaßen:

Sturzflugbremse	5	km/h
4 Außenliegende Bombenträger	14	"
Liegewanne für C-Stand	7	"
Eckige Scheiben zur Sichtverbesserung	5	"
Einbau des großen Rades mit großem Windabfluß	5	"
Flaches Monokel des B-Standes	4	"
Tarnanstrich	3	"
Zusatzantenne	2	"
	45	**km/h**

▨ Betriebsstoff
☐ Bomben

Comparison of the combat performances of the Ju 88 A and B.

Mock-up of the Ju 88 B cockpit.

Instrument layout in the new cockpit (mock-up).

The observer's position offered an excellent all-round view.

Ju 88 C-2 with additional armament of two MG 151 cannon.

The Ju 88 C-2 illustrated in the previous photograph with nose fairing removed.

Ju 88 B mock-up illustrating the excellent view offered the gunner in the second *B-Stand*.

As the war went on, the steady expansion of the war theater in Europe, North Africa, and over the Atlantic revealed that the *Luftwaffe* was stretched to the limits of its capabilities. The capacities of personnel and technology were inadequate to achieve air superiority in a single-front, then two- and three-front war. Increases in aircraft production could not keep up with combat losses, and the result was a "bottomless pit," in which men and machines were pushed far beyond their limits. The offensive *Luftwaffe* became a harried air force struggling in vain to maintain control of its own airspace.

The changing roles assigned to the Ju 88 and the new versions and variants developed to meet these roles provide some insight into the state of technology and organi-

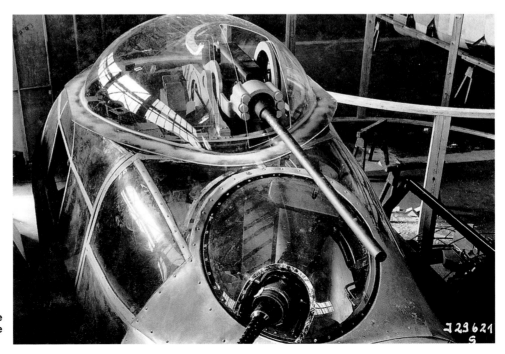

The *B-Stand-2* of the Ju 88 E. The power turret was armed with a single MG 151 cannon.

Double-barreled machine-gun in the ventral gun position of the Ju 88 B (mock-up).

A Ju 88 C-6 with upwards-firing armament of two MG 151/20 cannon. This installation, the so-called "Schräge Musik", was intended for use against heavy bombers.

zation in those days and illustrate the missions and problems of the front-line units. Increasing specialization in the development of the Ju 88 and the roles assigned to it reflected the multitude of strategic and tactical missions forced upon the *Luftwaffe*. It is therefore not surprising that a number of training versions had to be developed to provide a sound basis for pilots before entering combat. Each new specification and role resulted in a new version or variant of the Ju 88. The Ju 88 A-5 incorporated lessons learned in the Battle of Britain.

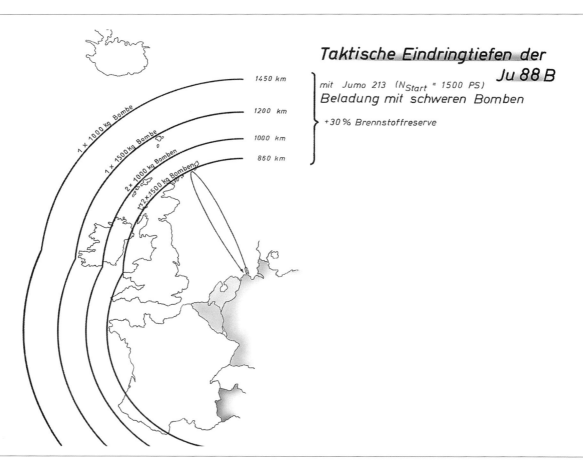

Taktische Eindringtiefen der *Ju 88 B*

mit Jumo 213 (N_{Start} = 1500 PS)
Beladung mit schweren Bomben

+30 % Brennstoffreserve

1450 km
1200 km
1000 km
860 km

1 x 1000 kg Bombe
1 x 1500 kg Bombe
2 x 1000 kg Bomben
2 x 1500 kg Bomben

Tactical penetration range of the Ju 88 B heavy fighter series powered by the Jumo 213.

Ju 88 A-5: Technical Specification

Airframe:

Fuselage: Monocoque fuselage with four longitudinal spars and bulkheads arranged at right angles to the longitudinal axis. Flush-riveted stressed skin exterior.

Undercarriage: Single-leg undercarriage members with double-brake wheels, which retracted rearwards hydraulically to lie flat beneath the wing. Emergency lowering hydraulic by hand pump over emergency line. Retractable tailwheel, capable of turning through 360°, lockable from cockpit hydraulically.

Control Surfaces:
a) Ailerons- Extending over 2/5 of the wing, metal covered with trim tabs and mass balance.
b) Landing Flaps- Extending over 3/5 of the wing, with flap locks, metal covered and with aerodynamic seal.
c) Horizontal Tail- Cantilever, stabilizer hydraulically coupled with landing flaps, metal covered, elevator mass balanced. Trim tabs assisted elevator when deflected, began dive and pull-out, hydraulic extension, trimming-back by spring tension.
d) Vertical Tail- Metal covered, rudder with trim tab, spring actuators and mass balance.

Wing: Two-spar wing, stressed skin.

Dive Brakes and Automatic Pull-Out Device: Hydraulically-activated dive-brakes on underside of outer wings. Automatic pull-out device hydraulically activated, coupled with bomb-release mechanism, appropriate valves allowed system to be operated with or without dive- brakes.

Heating and De-Icing: Safety air heating for crew compartment, leading edge heated by warm air, bladder de-icer on horiz. stabilizer, liquid de-icers on propellers, electric windscreen de-icer.

Hydraulic System: Hydraulic system operated the main undercarriage and tailwheel, landing flaps and horizontal stabilizer, dive-brakes and automatic pull-out system.

Power Plants

Engines: Two Jumo 211 B or G with mixture control, 1:1.68 reduction ratio, each producing 1,200 HP for takeoff at 2,400 rpm with 1.35 atm of boost.

Radiators: Annular radiators in front of each engine.

Propellers: Three-blade (in some cases four-blade) VDM variable-pitch metal propellers, 3.60 m diameter.

Fuel Tanks and Fuel:
Fully protected tanks distributed as follows:
One tank in forward bomb bay 1200 l
One tank in rear bomb bay 680 l
Two outer wing tanks 850 l
Two inner wing tanks 830 l
Total Fuel 3580 l
Non-consumable fuel, total of 45 l, fuel jettisoning for both fuselage tanks.

Recommended fuel: standard A 2 (octane number 87).

Oil Tanks, Oil and Oil Cooler: Two protected tanks in wings, each with a capacity of 136 l; maximum contents 2 x 125 l = 250 l. Also an unprotected tank in port wing with capacity of 105 l. Total oil capacity 355 l. Recommended lubricant: Rotring or Aero-Shell Medium. Oil Coolers: Air-tube cooler in upper part of each annular radiator.

Armament
A-Stand: 1 semi-rigid MG 15 with 450 rounds
B-Stand: 2 MG 15 in lens mounts and 1125 rounds
C-Stand: 1 MG 15, small lens mount and 600 rounds
Total: 2175 rounds

Bomb-Dropping Equipment and Bomb-Aiming Systems

External wing racks: four ETC 500/IXb or two Träger 1000 capable of accommodating four 500-kg or four 250-kg bombs, or one 1800-kg, two 1000-kg or two 500-kg bombs.

Aiming Devices:
For level bombing: Lotfe 7a or Lotfe 7b or BZG 2L or E or GV 219d.
For dive-bombing: BZA-1, Stuvi 5, Revi C 12c with swiveling plate.

Radio Equipment

Wireless	FuG X
Direction Finder	PeilG 5
Instrument Landing System	FuBl 1
Intercom	EiV

Dimensions, Weights and Performance

Dimensions and Weights:

Wingspan:	20.08 m (some A-5s 19.95 m)
Height:	4.85 m
Length:	14.36 m
Wheel Track:	5.80 m
Wing Area:	54.7 m²
Takeoff Weight:	13 000 kg

Maximum allowable landing weight (no bombs, fuselage tanks empty): 10 800 kg
Maximum allowable weight for unpowered or steep approach (with wing bombs and empty fuselage tanks): 10 700 kg
maximum wing loading: 240 kg/m²

Performance:

Maximum speed:
Outbound with flying weight of 13000 kg (with 2 x SD 1000 on external racks) at height of 5500 m   375 km/h
Return flight with flying weight of 10400 kg (bombs gone) at height of 5500 m 440 km/h

Maximum allowable speeds

50° dive (with dive brakes) at all heights (indicated airspeed)   575 km/h

20° to 30° dive (without brakes)

0 to 2000 meters	675 km/h
above 2000 meters	600 km/h
in bad weather at low altitude	400 km/h
with undercarriage lowered	250 km/h

Other
K 4ü autopilot with switch for single-engine flight

Takeoff-assist rocket fittings for takeoffs in overloaded condition.

Single-engine flight: just possible after release of bombs with empty fuselage tanks and external racks jettisoned (flying weight 10000 kg) at continuous power (2100 rpm with 1.10 atm of boost with supercharger in low blower ratio).

The above specification gives some indication of the bomber's versatility and performance, which, militarily and technically, was at the top of its class at that time. The Ju 88 A-5's qualities made it a vital component of the *Luftwaffe* and the type was fully comparable to its contemporaries. Deliveries of the Ju 88 A-4, which was equipped with more powerful Jumo 211 J engines, to the front-line units began in 1941, or after the Ju 88 A-5. The following data apply to the A-4 series.

Ju 88 A-4: Technical Specification

Airframe

Fuselage:	like the A-5
Undercarriage:	like the A-5, but with a retractable tailwheel with shock strut, also capable of turning through 360°, but self-centering and not lockable.
Control Surfaces:	a) slotted ailerons, trim tab on port aileron b) landing flaps: like the A-5 c) horizontal tail: like the A-5 d) vertical tail: like the A-5
Wing:	like the A-5
Dive Brakes and Automatic Pull-Out System:	like the A-5
Heating and De-Icing:	like the A-5
Hydraulic System:	like the A-5

Power Plants

Engines:	two Jumo 211 J with mixture control, 1:1.83 reduction ratio, producing 1420 HP for take off at 2600 rpm with 1.40 atm boost pressure Automatic switch from low to high supercharger at 3000 m.
Radiators:	Heavy metal annular radiators, later light metal annular radiators in front of each engine.
Propellers:	Fully-automatic, hydraulically-activated three-blade Junkers VS 11 wooden propellers. Diameter 3.6 m, selection range by means of rpm selector on throttle lever box between 1800 and 2600 rpm. Base setting 25°. Feathering electro-hydraulic.

Fuel Tanks and Fuel:	like the A-5
Oil Tanks, Oil and Oil Cooler:	like the A-5, but with heavy metal cooler

Armament

A-Stand:	1 x semi-rigid MG 81 with 750 rounds
B-Stand:	2 x MG 81 li, lens mounts with 1800 rounds
C-Stand:	1 x MG 81Z, small lens mount 1800 rounds
Total ammunition:	4,350 rounds

Bomb Dropping and Aiming Equipment
Two ETC 50/IXb (M 14) external bomb racks beneath the wings plus two Träger 1000 (M3) racks and the possibility of two additional ETC (500) in the aileron area capable of carrying: six 500-kg or six 250-kg bombs, or one 1800-kg, two 1400-kg, two 1000-kg or two 500-kg bombs

Aiming equipment:	like the A-5

Radio Equipment:	like the A-5 with addition of FuG 16 radio (together with FuG 10), FuG 25 IFF.

Other

Autopilot:	like the A-5
Single-Engine Flight:	like the A-5, however flying weight of 10500 kg possible at combat power (2500 rpm at 1.25 atm of boost, supercharger in low gear) with starboard engine running, 10300 kg with port engine running and in both cases radiator flaps of dead engine closed. In order to achieve this critical aircraft weight all bombs had to be dropped, all external racks jettisoned, both fuselage tanks emptied and the bomb sight and all non-essential guns, ammunition and armor jettisoned.

A comparison of the two series, the A-4 and A-5, shows the mature state of development the Ju 88 had reached by mid-1941. The more powerful Jumo 211 J engines made a significant contribution to this, improving maximum speed and handling characteristics. The aircraft's bomb load was also increased, while the number of external racks was increased to six.

The A-4 version of the Ju 88 was highly valued by the *Luftwaffe* on account of its excellent performance and good maneuverability in the horizontal and vertical planes. The aircraft often carried out steep diving attacks. The pilots of other bomber types reported vibration at the moment the bombs were released, however, this was rarely the case with the Ju 88. The Ju 88 also had a number of features which made it easier to control the aircraft; for example, the elevator could be automatically trimmed for the dive toward the target, or the exact reverse when the dive brakes were retracted after the pull-out. In each case the aircraft was automatically returned to level flight.

The Ju 88 A-5 was further developed, resulting in the Ju 88 B, with more powerful engines, a redesigned fully-glazed nose, and increased armament. A comparison between the Ju 88 A and Ju 88 B clearly shows the qualitative and design development of the aircraft and the evolution of the weapon in the interests of an offensive strategy. The various versions of the Ju 88 were characterized by their armament. The C and Z versions were heavy fighters and night fighters, the D and F were reconnaissance aircraft, the G, Q, and R versions were mainly night fighters, the H was a heavy fighter or long-range reconnaissance aircraft, the P version was equipped specially to engage enemy tanks and ground targets, the S version was a high-speed bomber and reconnaissance aircraft, and the T series was designed as a heavy fighter.

This multiplicity of variants and armament combinations ensured the optimal capability to engage air and ground targets and carry out tactical and strategic reconnaissance. The Ju 88's armament was selected according to its operational role, as in the following examples:

Ju 88 A-1, A-5 Bomber - 1938-40

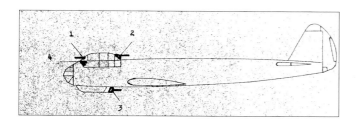

Defensive Armament:
1. *A-Stand*
   Mount: ball, diameter 100   Ikaria
   Weapon: 1 x MG 15   Rheinmetall-Borsig
   Ammunition: 375 rounds in five 75-round drums

2. *B-Stand*
   Mount: 2 x LLK lens mounts   Ikaria
   Weapon: 2 x MG 15
   Ammunition: 1,275 rounds in seventeen 75-round drums

3. *C-Stand*
   Mount: LLK lens mount   Ikaria
   later Bola 39 ventral mount   Rheinmetall-Borsig
   Weapon: 1 x MG 15
   Ammunition: 450 rounds in six 75-round drums

4. Sight
   Revi C 12 C reflector gunsight   Zeiss-Jena
   for fixed weapon in *A-Stand*

Offensive Armament:
1. Internal Loads
   (a)   Bomb Cell 1
   Racks: 2 racks, 4 x Schloss 50/X   comp. workshops
   (fuselage walls)   Neubrandenburg
   Bomb Load 18 x 50-kg = 900 kg
   (b) Bomb Cell 2
   Racks: 2 racks, 4 x Schloss 50/X   comp. workshops
   (fuselage walls)   Neubrandenburg
                1 framework, 2 x Schloss 50/X comp. workshops
   (fuselage center)   Neubrandenburg
   Bomb Load 10 x 50 = 500 kg
2. External Loads
                Load I
   Rack: ETC 50/X   SAM
   Bomb Load: max. 1000 kg
                Load II
   Rack: ETC 500/II   SAM
   Bomb Load: max. 500 kg
                Load III
   Rack: ETC 500/II   SAM
   Bomb Load: 250 kg
   Special case for Load I port side   Junkers comp.
                                        workshops
                                        Neubrandenburg
   Rack: rack with Schloss 1000
   Bomb Load: max 1800 kg
3. Sight for Diving Attacks
   Revi C 12 C with Sp 1 pivoting plate   Zeiss-Ikon
   later BZA bomb-aiming system
4. Sight for Horizontal Attacks
   Lotfe 0 7 bombsight   Zeiss-Jena
   or BZG 2 bombsight   Zeiss-Ikon

Ju 88 A-4 Bomber - 1940-44

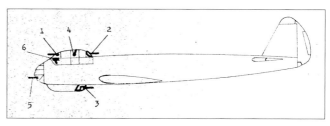

Defensive Armament:
1. *A-Stand*
   Mount: Lg 81 VE 45 B mount   Ikaria
   Weapon: MG 81   Mauser
   Ammunition: 750 rounds, belted

2. *B-Stand*
   Mount: 2 x LLK 81 VE lens mounts Ikaria
   Weapon: 2 x MG 81 I   Mauser

3. *C-Stand*
   Mount: Bola 81 z ventral mount   Ikaria
   Weapon: MG 81 Z
   Ammunition: 1000 rounds, belted, per gun

Ju 88 A-1, A-5 Bomber - 1938-1940

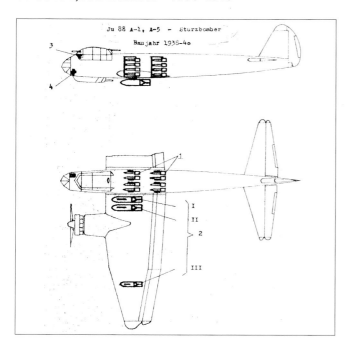

4. Additional Armament in Cockpit
Mounts:	2 ball mounts	Ikaria
Weapons:	2 x MG 15	Rheinmetall-Borsig
Ammunition:	450 rounds in 75-round drums	

5. Additional Armament in Nose
Mount:	Junkers	Junkers
Weapon:	MG FF/M	Ikaria
Ammunition:	120 rounds in two 60-round drums	

Ju 88 A-4 Bomber - 1940-1944

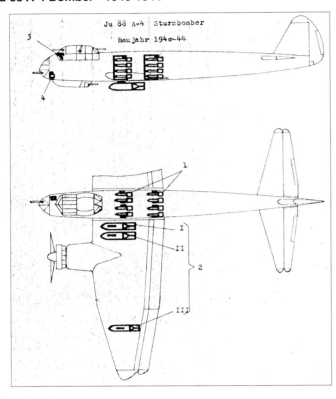

Offensive Armament:
1. Internal Loads
   Bomb Cell 1

Racks:	2 racks, 4 x Schloss 50/X (fuselage walls)	comp. workshops Neubrandenburg

   Bomb Load 18 x 50-kg = 900 kg

   (b) Bomb Cell 2

Racks:	2 racks, 4 x Schloss 50/X (fuselage walls)	comp. workshops Neubrandenburg
	1 framework, 2 x Schloss 50/X (fuselage center)	comp. workshops Neubrandenburg

   Bomb Load 10 x 50 = 500 kg

2. External Wing Loads (same both wings)
   Load I

Rack:	rack w Schlosslafette 500/1000	comp. workshops Neubrandenburg
Bomb Load:		max. 1 800 kg

   Load II
   Load III

Rack:	ETC 500/IX	SAM
Bomb Load:		500 kg
Special case for Load I port side		Junkers comp. workshops Neubrandenburg
Rack:	rack with Schloss 1000	Bomb Load: max 1800 kg

3. Sight for Diving Attacks   BZA 3 bombsight
4. Sight for Horizontal Attacks

Lotfe 7C bombsight	Zeiss-Jena
or Lotfe 7D bombsight	Zeiss-Jena

Ju 88 C-2 Heavy Fighter and Night Fighter - 1939-1941

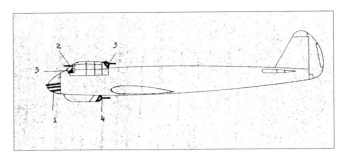

1. Fixed Armament
Mounts:	Junkers	
Weapons:	3 MG 17	Rheinmetall-Borsig
	1 MG 151/20 w/ StL 151/7	Mauser
Ammunition:	MG 17: 1,000 rounds per gun, belted	
	MG 151/20: 400 rounds, belted	

2. *A-Stand*
Mount:	ball mount	Ikaria
Weapon:	MG 15	Rheinmetall-Borsig
Ammunition:	375 rounds in five 75-round drums	

3. *B-Stand*
Mount:	2 x LLI lens mounts	Ikaria
Weapon:	2 x MG 15	Rheinmetall-Borsig
Ammunition:	1,350 rounds in eighteen 75-round drums	

4. *C-Stand*
Mount:	Bola 39 ventral mount	Ikaria
Weapon:	MG 15	Rheinmetall-Borsig
Ammunition:	525 rounds in seven 75-round drums	

5. Gunsight   Revi C 12 C

Ju 88 C-5 Radar-Equipped Night Fighter - 1944

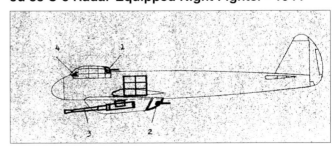

1. Mounts: F 1-A-103 Z   Rheinmetall-Borsig

2. | | | |
   |---|---|---|
   | Weapons: | 2 x MK 103 | Rheinmetall-Borsig |
   | Ammunition: | 200 rounds per gun, belted | |

3. A.I. Radar: Telefunken

4. *B-Stand*
Weapons:	2 x MG 81	Mauser
Ammunition:	900 rounds per gun, belted	
Mounts:	two LL-K 81 VE lens mounts	Ikaria

Ju 88 P-1 Tank-Destroyer - 1942

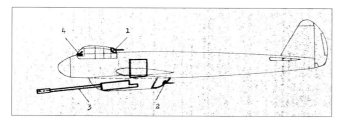

1. *B-Stand*
   Mounts: 2 x LL-K 81 VE lens mounts   Ikaria
   Weapons: 2 x MG 81   Mauser
   Ammunition: 900 rounds, belted, per weapon

2. *C-Stand*
   Mount: Junkers   Junkers
   Weapon: MG 81 Z   Mauser
   Ammunition: 800 rounds, belted, per gun

3. Fixed Armament
   Mount: Junkers   Junkers
   Weapon: Panzerkanone PK 40   Rheinmetall,
   Ammunition: 16 rounds   Dusseldorf
   Note: cocking semi-automatic (electro-pneumatic)

4. Sight   Revi C 12 C

Ju 88 P-3 Tank-Destroyer - 1943

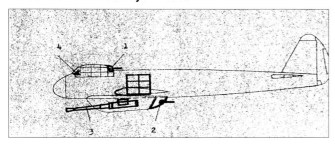

1. *B-Stand*
   Mounts: two LL-K 81 VE lens mounts   Ikaria
   Weapons: 2 x MG 81   Mauser
   Ammunition: 900 rounds, belted, per gun

2. *C-Stand*
   Mount: Junkers   Junkers
   Weapon: MG 81 Z   Mauser
   Ammunition: 800 rounds, belted, per gun

3. Fixed Armament
   Mount: Junkers   Junkers
   Weapon: 2 x 37-mm cannon   Rheinmetall,
   Ammunition: 90 rounds in 15   Dusseldorf
                6-round magazines

4. Sight   Revi 16 B   Zeiss-Jena

Ju 88 P-4 Tank-Destroyer (Variant A) - 1943

1. *B-Stand*
   Mounts: two LLK VE 81 lens mounts   Ikaria
   Weapons: 2 x MG 81 I   Mauser
   Ammunition: 900 rounds, belted, per gun

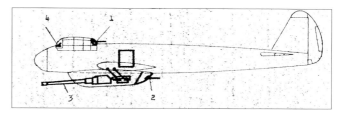

2. *C-Stand*
   Mounts: Junkers   Junkers
   Weapon: MG 81 Z   Mauser
   Ammunition: 800 rounds, belted, per gun

3. Fixed Armament
   Mount: Junkers   Junkers
   Weapon: 50-mm cannon   Rheinmetall,
   Ammunition: 42 rounds   Dusseldorf
   Note: cocking by hand

Ju 88 P-4 Tank-Destroyer (Variant B) - 1943

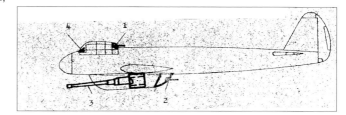

1. *B-Stand*
   Mounts: two LLK VE 81 lens mounts   Ikaria
   Weapons: 2 x MG 81 I   Mauser
   Ammunition: 900 rounds, belted, per gun
2. *C-Stand*
   Mounts: Junkers   Junkers
   Weapon: MG 81 Z   Mauser
   Ammunition: 800 rounds, belted, per gun
3. Fixed Armament
   Mount: Junkers   Junkers
   Weapon: 50-mm cannon   Rheinmetall,
                                               Dusseldorf
   Ammunition: approximately 36 rounds in drum magazines
   Note: automatic cocking, prototype only

Ju 88 A-17 Torpedo Bomber - 1942-1944

1. *A-Stand*
   Mount: Lg 81 VE 45 B mount   Ikaria
   Weapon: MG 81   Mauser
   Ammunition: 750 rounds, belted

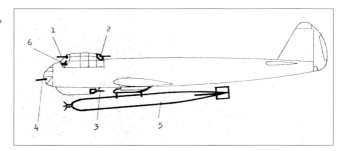

2. *B-Stand*
Mount:	2 x LLK 81 VE lens mounts	Ikaria
Weapon:	2 x MG 81 I	Mauser
Ammunition:	900 rounds per gun, belted	

3. *C-Stand*
Mount:	Bola 81 z ventral mount	Ikaria
Weapon:	MG 81 Z	
Ammunition:	900 rounds, belted, per gun	

4. Additional Armament in Cockpit Nose
Mounts:	Junkers	Junkers
Weapons:	MG FF/M	Ikaria
Ammunition:	180 rounds in two drum magazines, each 60 Rounds	

5. External Load
Racks:	2 x PVC 1006 B	MWN
Weapons:	2 x F5 or F5w torpedoes	

6. Sight   Revi C 12 C   Zeiss-Jena

Ju 88 with Rearwards-Firing Armament - 1941

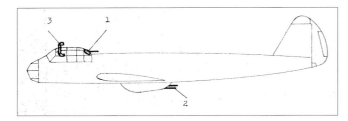

1. *B-Stand*
Mounts:	two LLK 81 VE lens mounts	Ikaria
Weapons:	2 x MG 81 I	
Ammunition:	750 rounds per gun, belted	

2. Fixed Armament
Mounts:	Junkers	Junkers
Weapons:	2 x MG 17	Ikaria
1 x MG FF		
Ammunition:	500 rounds per gun	
60 rounds n drum magazine		

3. Sight:
	RF I rear-view sight	Götz-Vienna

Ju 88 with Special Weapons Installation - 1942

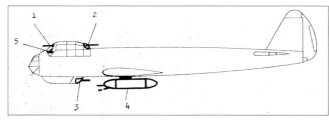

1. *A-Stand*
Mount:	Lg 81 VE 45 B mount	Ikaria
Weapon:	MG 81	Mauser
Ammunition:	750 rounds, belted	

2. *B-Stand*
Mount:	2 x LLK 81 VE lens mounts	Ikaria
Weapon:	2 x MG 81 i	Mauser
Ammunition:	900 rounds per gun, belted	

3. *C-Stand*
Mount:	Bola 81 z ventral mount	Ikaria
Weapon:	MG 81 Z	
Ammunition:	900 rounds, belted, per gun	

4. *Zusatzwaffenbehälter 81* (Additional Weapons Container)
   Junkers
Weapons:	6 x MG 81	Mauser
Ammunition:	200 rounds per gun, belted	
	or *Zusatzwaffenbehälter151*	Junkers
Weapons:	2 x MG 151	Mauser
Ammunition:	200 rounds per gun	

5. Sight   Revi C 12 D   Zeiss-Jena

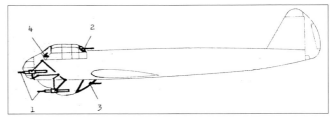

Ju 88 Special Installation (hand-served, semi-rigid 3-gun system) - 1941

1. *A-Stand*
Mounts:	Junkers	Junkers
Weapons:	1 x MG FF/M in nose	Ikaria
2 x MG FF/M in ventral gondola		
Ammunition:	360 rounds in six 60-round drums	

2. *B-Stand*
Mount:	2 x LLK 81 VE lens mounts	Ikaria
Weapon:	2 x MG 81 i	Mauser
Ammunition:	900 rounds per gun, belted	

3. *C-Stand*
Mount:	ball mount	Ikaria
Weapon:	MG 81 Zi	Mauser
Ammunition:	900 rounds, belted	

4. Sight   Revi C 12 C   Zeiss-Jena

Ju 88 *Düka*

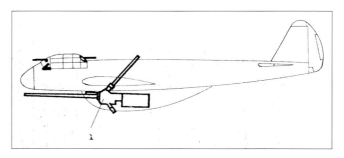

One 88-mm jet cannon
Did not enter service, firing tests only by Rheinmetall.

Ju 88 (with obliquely-mounted upwards-firing armament for use against heavy bombers)

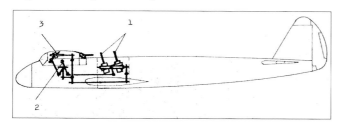

1. Oblique Armament
   Mounts: Rheinmetall-Borsig
   Weapons: 2 x MK 108   Rheinmetall-Borsig

2. Aiming Mechanism   Rheinmetall-Borsig

3. Sight   Revi 16 B   Zeiss-Jena

Ju 88 Mistel 2 (guided bomb) - 1944-1945

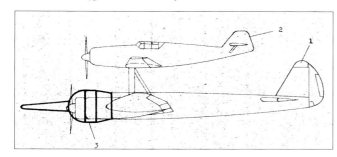

1. Ju 88 A-4 with Jumo 211 J

2. Bf 109 F-4 with DB 601 N

3. Warhead (4-ton)

Ju 88 with Reactive Armament (upwards-firing for use against airborne targets)

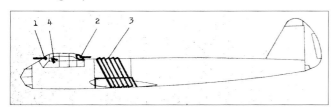

1. *A-Stand*
   Mount: Lg 81 VE 45 B mount   Ikaria
   Weapon: MG 81 i   Mauser
   Ammunition: 750 rounds per gun, belted
2. *B-Stand*
   Mount: 2 x LLK 81 VE lens mounts   Ikaria
   Weapon: 2 x MG 81 i   Mauser
   Ammunition: 900 rounds per gun, belted
3. Reactive Armament
   10-12 firing tubes
   Ammunition: 120-mm caliber (80 kg)
4. Sight   Revi 16 B
Note: The installation of reactive weapons was carried out only by the front-line units.

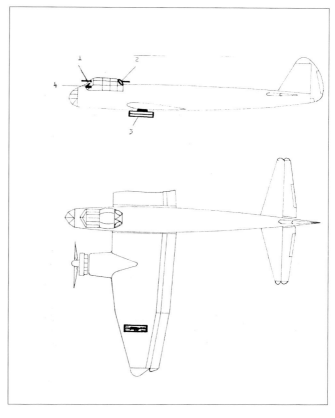

Ju 88 with Reactive Armament

1. *A-Stand*
   Mount: Lg 81 VE 45 B mount   Ikaria
   Weapon: MG 81 i   Mauser
   Ammunition: 750 rounds per gun, belted

2. *B-Stand*
   Mount: 2 x LLK 81 VE lens mounts   Ikaria
   Weapon: 2 x MG 81 i   Mauser
   Ammunition: 1,000 rounds per gun, belted

3. Reactive Armament
   (four launch tubes each side)
   210-mm caliber, weight 80 kg

4. Sight   Revi C 12 C

Ju 88 Reactive Armament - 1943

1. Drum with 11 rounds   Junkers
   Caliber 120-mm
   Weight 80 kg

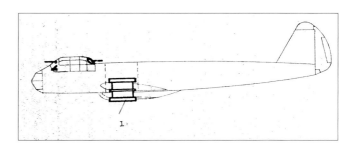

Note: This installation existed as a mockup only and did not see operational use.

Ventral Gondola Armament in the *C-Stand* position of the Ju 88 A-series with Bola 39 - 1928-1939

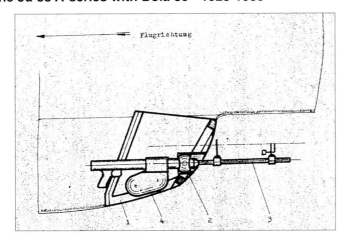

1. Jettisonable section (for emergency exit)   Junkers

2. Mount: LLK lens mount   Ikaria

3. Weapon: MG 15   Rheinmetall-Borsig

4. Spent shell casing bag

Ventral Gondola Armament in the *C-Stand* position of the Ju 88 with Bola 39 - 1939-1940

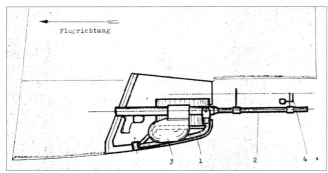

1. Bola 39 ventral mount   Ikaria
   (jettisonable in an emergency)

2. Weapon: MG 15   Rheinmetall-Borsig

3. Spent shell casing bag

4. Sight: wind vane front sight

Ventral Gondola Armament in the *C-Stand* position of the Ju 88 with Bola 81 Z - 1940-1944

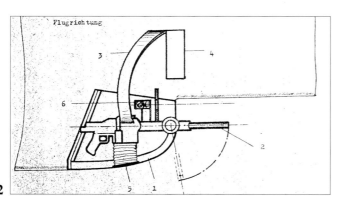

1. Bola 81 Z ventral mount   Ikaria
   (jettisonable in an emergency)

2. Weapon: MG81 Z   Mauser

3. Ammunition boxes

4. Spent shell casing chute

5. Sight: Revi 16 B

Ventral Gondola Armament in the *C-Stand* Position of the Ju 88 with Bola 131 - 1944

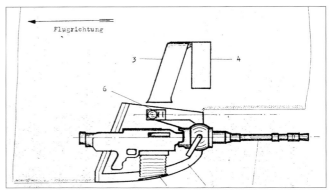

1. Bola 131 ventral mount   Ikaria
   (jettisonable in an emergency)

2. Weapon: MG 131

3. Ammunition feed

4. Ammunition boxes

5. Spent shell casing chute

6. Sight: Revi 16 B

The Following Companies or their Licensees Delivered Weapons or Weapons Related Equipment to Junkers for Installation in the Ju 88:

	Company	Address	Products
1.	AEG	Berlin	remote-control systems
2.	Arga, formerly LAB or Argus	Berlin	fixed mounts, flexible mounts, remotely-controlled mounts
3.	Askania	Berlin	gun sights
4.	Auto-Union development	Chemnitz	mount conversion and
5.	Görz	Vienna	bomb sights, periscopes
6.	Kurt Heber	Osterode	rocket launching equipment
7.	HoList systems	Berlin	automatic bomb-release
8.	Ikaria	Berlin	aircraft cannon under license, fixed mounts, flexible mounts
9.	Kickert	Dusseldorf	bomb-dropping equipment
10.	Maihag	Hamburg	pneumatic cocking equipment for machine-guns
11.	Mauser	Oberndorf	machine-guns, cannon
12.	Mechanische Werkstätten (MWN)	Neubrandenburg	bomb- and torpedo-dropping equipment
13.	Michel	Augsburg	weapons switchboxes
14.	Rheinmetall-Borsig MWK	Berlin, later Celle District	machine-guns, cannon, flexible mounts, remote-control mounts
15.	Rheinmetall Sömmerda	Sömmerda	electrical bomb-fusing systems
16.	Siemens LGW	Berlin	bomb-dropping and electric gun cocking systems
17.	Steinheil	Munich	periscopes, gun sights
18.	Telefunken	Berlin	gunnery radars
19.	Zeiss	Jena	bomb sights, gun sights, periscopes
20.	Zeiss-Ikon	Dresden	bomb sights

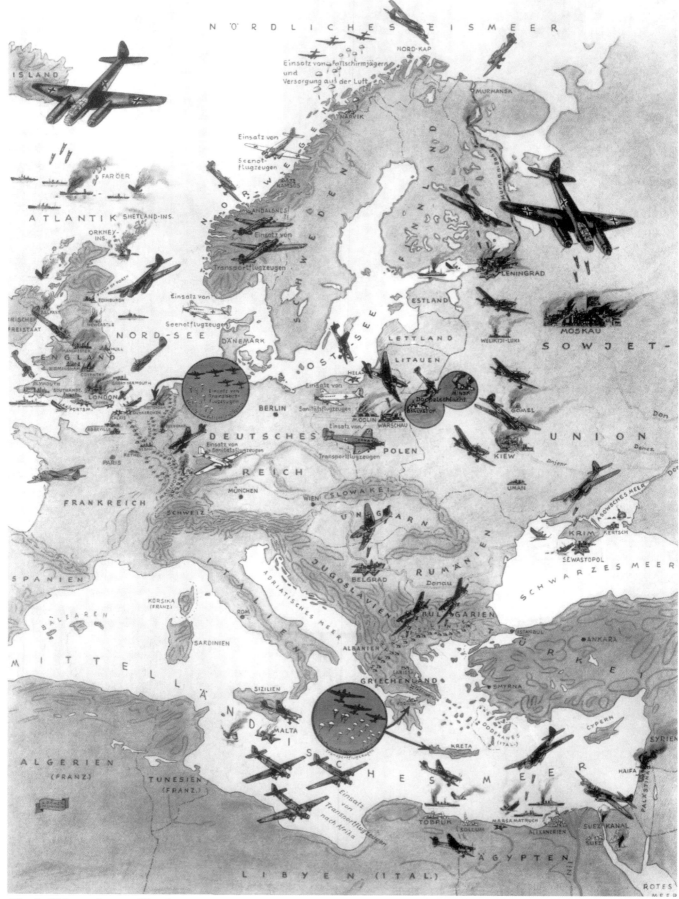

The Ju 88 in service on all fronts.

Flying Bomb:
The Ju 88 *Mistel*

The so-called "Mistel Combination" was probably the most bizarre development in the Ju 88's story. In 1943 Junkers (at the request of the *Reichsluftfahrtministerium*) began developing a so-called "piggybacked" *Mistel*, or "Father and Son" aircraft. The concept envisioned a Ju 88 converted to serve as a flying bomb with a command and control aircraft riding atop a system of struts. It was anticipated that obsolete or irreparable Ju 88s would be fitted with an explosive warhead. The command aircraft would guide the combination to the target, where the Ju 88 would be released.

As the two aircraft were joined by struts, separation was achieved by explosive bolts. The intended targets were Allied ships in the west and Soviet armament factories and the Oder bridges in the east. One pilot, *Oberleutnant* Vollhard of *3. Staffel/NJG 7*, claimed to have carried out an attack against the Oder bridges on 20 April 1945.

Such operations were risky ventures, as the clumsy combinations were easy prey for Allied fighters. Only rarely did a Ju 88 flying bomb hit its assigned target.

By 1944 large parts of the Dessau works had been evacuated to the Harz Region. Consequently, most of the Mistel combinations were assembled in the Nordhausen area and the eastern Harz. Various sources suggest that four, possibly five, versions were built, with production totaling more than 100 piggyback aircraft. Another variant, a combination of a Ju 188 and a Me 262 control aircraft, failed to reach the test stage.

Ju 88 Mistel

Mistel 1	Ju 88 A-4 with Bf 109 F-4
Mistel S 1	Ju 88 A-4 with Bf 109 F-4 (training variant)
Mistel 2	Ju 88 G-1 with Fw 190 A-6
Mistel 3 A	Ju 88 G-10 with Fw 190 F-8
Mistel 3 B	Ju 88 H-4 with Fw 190 A-8
Mistel 3 C	Ju 88 H-4 (hollow-charge warhead installed in nose) with Fw 190 A-8
Mistel 4	Ju 188 with Me 262 (planned)
Mistel 5	Ju 88 G-7 with Ta 152 H (flight tests, no operations)

A Mistel-2 combination, consisting of a Ju 88 G-1 and a Fw 190 A-6.

Photographed after capture by the British: a Mistel-2 combination.

For the Airplane Modeller

Along with information on the most important technical and tactical data of the Ju 88, the airplane model builder will also find a series of drawings and phptographs, from which he can find a picture for modelmaking. Naturally, the book at hand is not a thorough work on a specific subject. It should and could not be such, for along with the series and variations of the Ju 88 and the additional and sometimes also individual modifications at the front, there exists such a vast array of possibilities for this aircraft type that they cannot be examined in depth on account of the thematic brevity. Detailed aspects would have to be reserved for a larger monograph. In this book, there should also be further attention paid to the numerous standard kits that are available in the trade. This is also made clear by the included model photos and illustrations. Photos of models made by young people twelve to fourteen years old have been chosen. This should provide encouragement for young people to take an interest in model building and get into the hobby. At such an age they can quickly develop practical abilities and experience. Then too, there is a distinct possibility of getting especially involved with history, particularly military and aeronautical history. And the last two pictures come directly from Junkers material.

Model of a Ju 88 A-4 with camouflage paint.

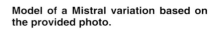

Model of a Mistral variation based on the provided photo.

Model of the Ju 88 A-4 with bomb load.

**Model of the Ju 88
B Series.**

Bilbliography

Bornemann, M.: Geheimprojekt Mittelbau, Bonn 1994
Bracke, G.: Melitta Gr5fin von Staufenberg -
Das Leben einer Fliegerin, München 1990
Brausewaldt, H.: Sturzkampfflugzeuge, Berlin/Leipzig 1941
Bukowsky/Griehl: Junkers-Flugzeuge 1933-1945, Friedberg 1991
Cescotti, R.: Kampfflugzeuge und Aufkldrer, Koblenz 1989
Dressel/Griehl: Deutsche Sturzkampfflugzeuge Ju 87, Ju 88; Friedberg 1992
Feuchter, G. W.: Der Luftkrieg, Frankfurt/ Bonn 1964
Filley, B.: Junkers Ju 88 in action, Part 1, Carrolton 1988
Filley, B.: Junkers Ju 88 in action, Part 2, Carrolton 1991
v. Gersdorff/Grasmann/

Schubert: Flugmotoren und Strahltriebwerke, Bonn 1995
Griehl, M.: Junkers-Bombers, Dorset 1987
Griehl, M.: Junkers Ju 88, Friedberg 1989
Griehl, M.: Junkers Ju 88, Star of the Luftwaffe; London 1990
Groehler, 0.: Geschichte des Luftkrieges, Berlin 1981
Groehler, 0.: Kampf um die Luftherrschaft, Berlin 1988
Groehler, 0.: Anhalt im Luftkrieg, Dessau 1993
Held, W.: Reichsverteidigung - die deutsche Treibjagd 1943-1945; Friedberg 1988
Hentschel, G.: Die geheimen Konferenzen des General-Luft zeugmeisters, Koblenz 1989
Irving, D.: Die Trag6die der deutschen Luftwaffe, Frank furt/Berlin 1970

Junkers-Flugzeug- und Motorenwerke
Dessau: - Akten im Deutschen Museum München, 1935-1945 Ju 88 betreffend
- Bedienungsvorschriften der Ju 88 von 1941-1944
- Der Propeller, Werkzeitung der JFM, Jahrgang 1940 ff.
- Junkers-Nachrichten, Jahrgang 1940 ff.
- Lehrbldtter ftir die Monteurschulung Ju 88 betreffend
Kollektiv: Die Luftwaffe, Eltville 1993
Kollektiv: Luftkriegfiihrung im Zweiten Weltkrieg, Herford/Borm 1993
Kollektiv: Junkers Ju 88, Tokio 1995
Kosin, R.: Die Entwicklung der deutschen Jagdflugzeuge, Koblenz 1990
Lange, B.: Typenhandbuch der dentschen Luftfahrttech nik, Koblenz 1986
Laschoter, J.: Mit Ju 88 gegen England, Wien 1942
Ladwoch, J.: Junkers Ju 88, Warschau 1992

v. Medem, W. E.: Fliegende Front, Berlin 1942

Neitzel, S.: Der Einsatz der deutschen Luftwaffe ijber dern Atlantik und der Nordsee, Bonn 1995
Nowarra, H.J.: Die Ju 88 und ihre Folgemuster, Stuttgart 1987
Perlia, R.: In geheimer Mission, Illertissen 1996
Todte, H: Die Junkerswerke in Bernburg (Artikelserie in Bernburger Zeitung), Bernburg 1993/1994
Schmitt, G.: Junkers-Bilderatlas aller Flugzeugtypen 1910 1945, Berlin 1990
Schmitt, G.: Das Junkers-Flugzeugtypenbuch, Dessau 1997
Wagner, W.: Hugo Junkers - Pionier der Luftfahrt, Bonn 1996

Ju 88 in the annex of the USAF Museum in Dayton.

Ju 88 R-1, coded D5+EV, Werk.Nr. 360043, in the RAF Museum, London.